Lead and Public Health:
the dangers for children

Erik Millstone

Taylor & Francis
Publishers since 1798

Acknowledgement

I am grateful to John Russell for sharing the results of his researches with me.

First published in the US 1997 by
Taylor & Francis
1101 Vermont Avenue, NW
Suite 200
Washington DC 20005-3521
Tel: 202 289 2174   Fax: 202 289 3665

First published in the UK in 1997 by
Earthscan Publications Ltd

ISBN: 1 56032 723 5 (hardback)   1 56032 724 3 (paperback)

Typesetting and page design by PCS Mapping & DTP,
Newcastle upon Tyne, UK

Printed and bound in England by
Biddles Ltd, Guildford and Kings Lynn

Library of Congress Cataloguing in Publication data available from
Taylor & Francis

Table of Contents

List of Figures

List of Tables

Abbreviations and Acronymns

ATSDR	Agency for Toxic Substances and Disease Registry (US)
CDC	Centre for Disease Control (US)
CLEAR	Campaign for Lead-Free Air (UK)
DoE	Department of the Environment (UK)
EPA	Environmental Protection Agency (US)
GCI	General Cognitive Index
HUD	Department of Housing and Urban Development (US)
ILZRO	International Lead Zinc Research Organization (US)
IPCS	International Programme on Chemical Safety
IQ	intelligence quotient
K-TEA	Kaufman Test of Educational Achievement
LBP	Lead-based Paint
LDA	Lead Development Association (UK)
LIA	Lead Industries Association (US)
MDI	Mental Development Index
MRC	Medical Research Council (UK)
MSCA	McCarthy Scales of Children's Abilities
NAAQS	National Ambient Air Quality Standard
NHANES	National Health and Nutrition Examination Survey (US)
NSF	National Sanitation Foundation
OFWAT	UK Office of Water Regulation
PbB	blood lead level
PbT	tooth lead level
RCEP	Royal Commission on Environmental Pollution
UBK	Uptake/Biokinetic Model (US)
US FDA	US Food and Drug Administration
WHO	World Health Organisation
WISC	Wechsler Test of Child Intelligence
WISC-R	Wechsler Revised Intelligence Scale for Children
WRC	Water Research Centre (UK)

Lead: the Threat to Human Health

L ouis Sullivan, then US Secretary of Health and Human Welfare, stated on 7 October 1991 that 'Lead poisoning...is the number one environmental threat to the health of children.'[1] Trying to rank the importance of different toxic hazards faced in our communities is not a particularly precise science, but the case against lead is overwhelming: not only is it extremely poisonous, but a significant portion of the people living are carrying lead loads that are doing them harm. Sullivan was responding to a crucial report from the US Centre for Disease Control (CDC) in Atlanta entitled *Preventing Lead Poisoning in Young Children: A Statement by the Centre for Disease Control.*[2] The CDC had gone slightly further than Sullivan, asserting not just that 'Lead poisoning is one of the most common pediatric health problems in the United States today', but that '... it is entirely preventable.'[3] That latter assertion is important because it partly explains why the issue of lead pollution has gained such a high position on the environmental policy agenda in the US, despite the fact that it may not be the most ubiquitous environmental contaminant, a dubious privilege which is probably held by sulphur dioxide or by the plasticizer di–2–ethylhexyl phthalate (more commonly known as DEHP).

One reason why lead pollution is so high on the American policy agenda is because scientists have been able to produce convincing evidence of the harm which lead is causing, but secondly because that harm is adversely affecting the mental development of young children, and in a society which purports to believe, if not in the equality of economic outcomes, then at least in the equality of opportunity, damage to the life chances of innocent young children is unacceptable. A third reason is that, technically speaking, the problem can quite easily be addressed. In the US the problem has been recognized, and something is being done about it. In the UK, on the other hand, and for reasons which deserve an explanation, the problem is not acknowledged and little is being done to resolve it.

One reason why the problem of childhood lead poisoning is not being recognized in the UK is because young children in Britain are not being screened, whereas in the US there is a targeted childhood blood lead screening programme. Most of the evidence which would establish the scale of the problem in the UK has never been collected, and much of the information which has been collected remains unpublished. Sufficient evidence is available, however, to enable us to conclude that the problem of lead pollution in Britain is at least as serious as that in the US. There is at least as much old leaded paint in the British housing stock as there is in American homes, and a far higher propor-tion of British homes receive their drinking water through lead pipework than is the case in the US. In 1993 at least 30 million, ie some 12 per cent, out of approximately 250 million American citizens were receiving drinking water which contained at least 15 microgrammes of lead per Litre (µg/L), and approximately 4 million homes had lead-based paint in a hazardously deterio-rated condition. At the same time, a larger fraction of all British households, approximately one quarter, had their drinking water contaminated at levels above 10 µg/L. The net result has been that in both the US and the UK, over the last ten years, approximately 10 per cent of young children have been carrying levels of lead contamination sufficient to cause evident damage to their mental performance and development.

When in 1991 Jane Lin-Fu, then Director of the Childhood Lead Poisoning Prevention Program of the US Department of Health and Human Services, characterized lead poisoning as '...an ancient disease uniquely neglected by modern medicine...'[4] she was exaggerating, if only slightly. Lead poisoning has not been entirely neglected by the medical profession, and even if it had, that would not have been unique. In practice, a small but diligent group of scientists have been actively investigating the issue and their research has been fruitful. They have been able to demonstrate adverse effects of lead poisoning at ever lower levels of exposure.

Lin Fu is right that the hazards from lead have long been recognized. There are indications that both Hippocrates in the fifth century BC, and Galen in the second century BC, recognized that slaves working in lead mines suffered consequent neurological and kidney damage. Benjamin Franklin observed in 1786 that printers who worked with molten lead and cast their own type were suffering as a consequence, and in 1904 a doctor in Queensland, Australia, reported that young children were suffering with lead poisoning as a consequence of their exposure to contaminated paintwork. The evidence showing that lead is hazardous has continued to accumulate and strengthen, and in particular the levels of contamination down to which adverse effects have been detected has continued to decline. Body lead loads have been declining in both the US and the UK, and in other parts of the industrialized world too, but the blood lead levels at which adverse effects have been found have been falling no less rapidly. Responsibility for neglecting the problem of lead poisoning, and especially the damage which is being inflicted on young children, must be borne not by scientists but by governments.

There have, however, been important differences between the responses of the governments of the UK and the US. During the 1970s and early 1980s both governments were passive and complacent on this subject. Since 1986, however, the US government has been taking the problem of lead pollution seriously, and the contrast between the responses of the American and British governments since then has been stark. How the scientific evidence emerged, and how our two governments – in the US and the UK – have responded, or failed to respond, is the subject of this book.

The change in 1986 was accomplished by the publication of what, in its day, was the most thorough and comprehensive toxicological assessment ever produced.[5] The team of officials which prepared the four-volume document was under the leadership of a group of very senior scientists at the Environmental Protection Agency (EPA). Their efforts galvanized the American government. The response of ministers in the British government to the publication of that landmark report was to pretend that it had never emerged, and to try to ensure that no one's attention was drawn to it. If the document, or the evidence on which it was based, had been acknowledged in the UK there would have been pressure for a regulatory response, but Margaret Thatcher had made it clear to all her ministers and ministries that she was determined, as far as possible, to dismantle regulations, not to strengthen them.

It would be easy to provide a list of all of those parts of the human body which lead has been shown to damage; the problem is that the list will be an extremely long one. It might be simpler to list those few parts in which harm from lead has

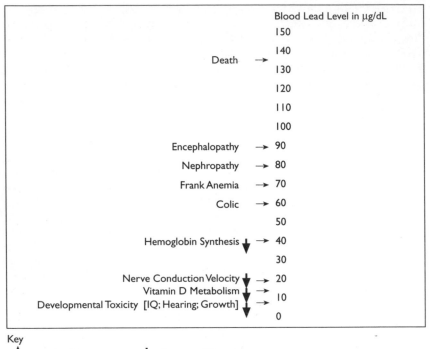

Key

↑ Increased Function   ↓ Decreased Function

*The levels in this diagram do not necessarily indicate the lowest levels at which lead exerts an effect These are the levels at which studies have adequately demonstrated an effect.

Source: Centre for Disease Control (1991) *Preventing Lead Poisoning in Young Children: A Statement by the Centre for Disease Control*, CDC, Atlanta

Figure 1.1: *Lowest Observed Effect Levels of Inorganic Lead in Children**

not yet been found. Figure 1.1 provides an illustration which shows the various levels of lead in the blood at which several specific adverse effects have been found in children. Blood lead levels are usually measured in terms of the number of micrograms (µg) of lead per 100 millilitres (ml) of blood and 100 ml is usually abbreviated to a decilitre (dL) (or one tenth of a litre). The illustration shows the blood lead levels for selected adverse effects ranging from damage to mental performance and development, growth and hearing, which occur at levels of 10 µg/dL or lower, to a significant reduction in the synthesis of haemoglobin at levels of approximately 40 µg/dL, kidney damage above 60 µg/dL, and to fatal lead poisoning in children as their blood lead levels approach 135 µg/dL.

While lead damages many parts of the body, it has been the effects of lead upon the brain, and the damage which it might cause to our ability to think and to function cognitively, that have been the main focus of the toxicological debate. Louis Sullivan and Jane Lin-Fu are entitled to rank lead poisoning at the top of their list of environmental poisons for at least three reasons. Firstly, because lead inflicts its greatest harm upon those most vulnerable, namely babies and young children; secondly, because the harm it does persists and weakens their abilities to cope with the remainder of their lives.[6] Thirdly, when scientists discuss the possible adverse effects of environmental contaminants on human beings, for example when they talk about dioxins, they are usually trapped in the tricky business of trying to guess what effects those compounds might have on humans, by drawing inferences from what has been learnt about their effects on laboratory animals. Studying the toxicity of lead has been unusually straightforward because much of the crucial evidence comes from studies of people, rather than of rats and mice. As the CDC puts it, the hazard is not just a theoretical one.[7] For at least the last 30 years, the debates have not been about whether or not lead is poisonous, but about:

- the levels at which adverse effects can be detected;
- whether or not there are thresholds below which those effects cease to occur; and
- whether or not the results of over-exposure persist or are merely transient.

While the evidence of the toxicity of lead at ever lower levels of exposure has continued to emerge, no evidence has emerged suggesting that there is a threshold below which adverse effects cease to occur, nor is there any evidence that lead performs an essential or even useful biochemical function in the human body.

Those considerations have, of course, had a direct bearing on the policy questions which governments have had to confront. Governments need to decide how large a margin of safety they should allow between body lead loads and levels at which proven adverse effects appear, and they need to consider the vital question of who should pay to clean the lead out of our environment. When it comes to the question of who should pay, there are normally just four candidates for the position: the lead industry, the water companies, the taxpayers or the owners of lead-contaminated property.

It is hard to decide how the responsibility for dealing with the problem should be shared because most instances of lead poisoning are not a consequence of modern industrial activity, but of historical uses of lead. Children are being poisoned mainly because of the presence of the old leaded paint with which their homes were decorated in years gone by, and because their water supply is reaching them through old and totally outdated water pipes. It is far easier to oblige an industry to reduce the extent to which it pollutes our environment than it is to clear up the mess when those responsible are no longer around to clean it up themselves.

If lead was not really a problem – or if it were, but everything that could and should be done *was* being done to solve it – there would be little point in my writing this book, nor in your reading it. Unfortunately, lead pollution *is* a serious problem. Technically it would be a relatively straightforward problem to solve, but it would not be cheap. Much that could and should readily be done, however, is not being done, and therefore this book is intended to be of practical, and not merely theoretical, relevance to its readership.

The election in Britain of a new Labour government in May 1997 moreover created a fresh opportunity. Although it is unlikely that substantial amounts of extra public money will flow into decontamination projects, the new government indicated that it was committed to preventative health measures. It has also emphasised its intention to raise educational standards amongst young children. The government may therefore be keen to reduce forms of pollution which can be shown to be damaging the educational attainment of this vulnerable group.

There remains, however, one important lead-related health issue which this book does not directly address and that concerns the exposure of adults to lead in the course of their employment: what officials and scientists call 'occupational exposure'. This book focuses rather narrowly on the hazards to young children in the general population. The protection of adult occupational health is also very important, but it remains beyond the scope of this text. This is partly because in relation to adults, the toxicological science is very much more complicated, and partly because it is governed by entirely separate institutions both in the US and the UK. The subject is sufficiently important, complex and interesting, however, for it to deserve an entire book to itself.

2

Is Lead Poisonous?

When we want to understand the hazard which lead pollution can pose, we have to turn to the science of toxicology. Scientific evidence can never by itself be sufficient to decide public policy, but without a clear grasp of what the scientists have found it is impossible to decide how far we should go to diminish our exposure to lead.

It is often alleged that economics is a 'dismal' science, by which people mean that economic theory is too poor to provide either reliable forecasts or an adequate explanation of processes of change.[1] It is, however, less common to find the science of toxicology being characterized in similar terms.[2] A full account of the general characteristics and limitations of toxicology is beyond the scope of this discussion, but as the debates about the toxic hazards from lead are reviewed it will be impossible to avoid acknowledging that some of the evidence is equivocal, none of the studies is definitive, that the interpretation of results is frequently contested, and that there is very little about which unanimity has been achieved.

It would be a serious mistake, however, to suppose that all studies are equally unreliable. During the last 25 years, toxicologists have witnessed striking improvements in the sensitivity and precision with which it has been possible to detect and measure both body lead loads and the adverse effects which lead causes. Not only has evidence emerged of adverse effects at progressively lower and lower levels, but also the reliability of that evidence has progressively strengthened. This branch of toxicology has been markedly less dismal than that relating for example to saccharin, where as evidence has accumulated, uncertainty has increased rather than diminished.[3] Even though every study has its limitations, and the strengths and limitations of each study deserve to be acknowledged, the reliability and conclusiveness of the evidence has improved with time, and not only should each study be judged on its merits, but the cumulative weight of the entire body of evidence also deserves careful consideration.

What are the Toxic Effects?

It is very easy to ask the simple question: is lead poisonous? To provide an adequate answer is rather more complicated. Any sensible response must start from a recognition that the damage which lead can inflict depends on the level of exposure, and the characteristics of the people being exposed – young children being especially vulnerable. One point upon which all competent protagonists in the lead poisoning debate are agreed is that if a person has high levels of lead in their blood, toxic effects are unavoidable. There is some consensus about what counts as a fatally high level, but the lower the level of exposure the sharper the disputes have been.

The total volume of literature concerning the possible toxicity of lead is monumental. It is so large that it is doubtful if there is any scientist in the world who has read, let alone remembered, all of it. In 1817 Orfilia observed that 'If we were to judge of the interest excited by any medical subject by the number of writings to which it has given birth, we could not but regard the poisoning by lead as the most important ... of all those that have been treated of, up to the present time.'[4] Little appears to have changed, in that respect at any rate, during the intervening years. When the British government received an official report in 1980 entitled *Lead and Health,* the authors said they had identified a bibliography of nearly 3000 relevant publications.[5] A huge number of new studies have been published since 1980. The review which this book provides will necessarily therefore be a selective account. The criterion by which the material has been selected, however, deserves to be acknowledged. The touchstone by which I have selected the elements of the scientific debate to include or omit has turned on the question of whether or not they have influenced, or deserve to have influenced, policy-makers in either the UK or the US.

Where does lead cause damage?

The human body is a very complex organism. Part of the complexity lies in the fact that it is both highly differentiated and highly integrated, that is to say there are lots of different parts, but they function most effectively when they all support each other. Although lead is toxic, it does not distribute itself evenly throughout the body, and it does not cause equally serious or similar types of damage to all kinds of tissues and organ sites. Lead is a cumulative poison which can be absorbed far more rapidly than it can be excreted, and lead preferentially accumulates in hard tissue such as bone. The illustration in Figure 2.1 provides a simplified but useful representation of some of the ways in which lead comes to be differentially located in some compartments of our bodies.[6]

Even though lead may exert toxic effects in many parts of the body, what matters for this discussion is not an exhaustive account of all possible adverse effects, but rather the identification of the most vulnerable parts of the body, the groups most vulnerable to having those systems damaged, and an analysis of how they can most effectively be protected.

How body lead loads can be estimated

When we want to estimate the levels of lead being carried by particular individuals or groups we encounter a tricky problem. There is no single indicator which can fully represent the lead burden in a person's body. We have good reasons for thinking that, within any person's body, the levels of lead in the bone will differ from those in their kidneys, brain or blood. We should also expect those relationships to vary from time to time and from person to person. In toxicological terms, the most important levels are those in the tissues in which lead is causing the most serious adverse effects; that in turn raises a problem which is not entirely insuperable, but is uncomfortably close to being so. The most heated controversy in lead toxicology has focused on the damaging effects which lead has on the human brain. It is, however, totally unacceptable directly to intrude into the brains of living humans for experimental purposes. Despite the development of various kinds of brain scanners, we do not yet have a non-invasive method with which to estimate levels of lead in the brain.

In practice, scientists have concluded that until an acceptable non-invasive technique becomes available, the best alternative is provided by measuring levels of lead in the blood,

Inhaled and ingested lead circulates via blood (1) to mineralizing tissues such as teeth and bone (2), where long-term retention occurs reflective of cumulative past exposures. Concentrations of lead in blood circulating to 'soft tissue' target organs such as brain (3), peripheral nerve and kidney reflect both recent external exposures and lead re-circulated from internal reservoirs (eg bone). Blood lead levels used to index internal body lead burden tend to be in equilibrium with lead concentrations in soft tissues and, below 30µg/dl, also generally appear to reflect accumulated lead stores. However, somewhat more elevated current blood lead levels may 'mask' potentially more toxic elevations of retained lead due to relatively rapid declines in blood lead in response to decreased external exposure. Thus, provocative chelation of some children with blood leads of 30–40 µg/dl, for example, results in mobilization of lead from bone and other tissues into blood and movement of the lead (4) into kidney (5), where it is filtered into urine and excreted (6) at concentrations more typical of overtly lead-intoxicated children with higher blood lead concentrations.

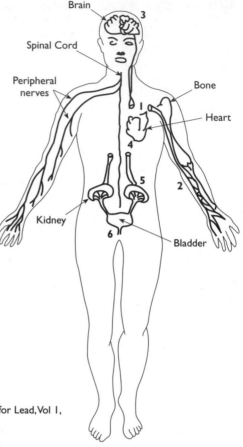

Source: EPA (1986) Air Quality Criteria for Lead, Vol 1, p 1-134.

Figure 2.1: *Illustration of some of the main compartments of the body in which lead may have a significant presence*

although it is important to bear in mind the limitations of such measurements. When a specialist committee of the World Health Organisation discussed this matter in 1992, they reported that 'There is a lack of reliable data to develop any scientific position regarding the "active" fraction of lead in blood...the relationship between blood lead in and brain lead in developing children, or the impact of transfer of lead into the brain.'[7] Very similar comments apply to our knowledge of the effects which lead has on kidneys and bones. Our knowledge, therefore, of the quantities of lead present in the most important parts of people's bodies is therefore indirect and inferential, and consequently needs to be interpreted with care.[8]

Scientists can only scrutinize those bodily materials which they can readily, safely and acceptably obtain. When trying to estimate human exposures to lead, scientists have confined their attention mainly to teeth, hair and blood. For reasons which will shortly be explained, scientists have concluded that, for the time being at any rate, it is the levels of lead in the blood which provide the best indicator of a person's body lead load. Consequently when body lead loads are discussed in both the scientific literature and in policy debates, the discussion typically uses levels of lead in the blood as the primary indicator of body lead loads. The phrase 'blood lead level' is used so frequently that it has come to be abbreviated to the expression 'PbB'.

Use of blood-lead measurements

One of the main problems with measuring the body's lead load in terms of the quantity of lead in a given volume of blood is that it provides a reliable indication of only relatively recent exposures. Estimates of the rate at which blood lead levels can change vary, but some commentators estimate that blood lead levels (PbB measurements) accurately reflect, at most, exposure during the preceding month.[9] Cory Schlechta found that the same PbB level could be found in rats exposed to lead for 1, 8 or 11 months, although using different dosing regimes; but even more significantly, the same PbB levels resulted in quite different behavioural expressions.[10] Consequently, a single blood lead reading may be an adequate indicator of very recent exposure, but it will be a poorer indicator of historic exposures and of residual accumulations.[11] The levels of historical exposure and residual accumulations may, however, be precisely those which are toxicologically most significant, although they can be rather hard to estimate. In recent years scientists have tried to devise a means for coping with that problem.

The scientists' initial motivation was to deal with a rather separate aspect of their difficulties. For several years evidence had been available which showed that there is a close and inverse relationship between blood lead levels and mental performance, as estimated using IQ tests, especially in children. Those wishing to defend the lead industry against scientific criticism, and from the imposition of official controls, repeatedly argue that simply finding correlations between PbB measurements and scores on IQ tests did not prove that higher lead levels were damaging children's brains. It might merely be the case that children who are less bright are more likely to be ingesting lead than the smarter children. They might, for example, be more likely to put into their mouths toys on which dust contaminated with lead had gathered.[12] Those scientists who strongly suspected that lead is the cause of neurological damage saw that the most effective response to that line of argument would be to conduct what are called longitudinal prospective studies. In a longitudinal and prospective study, estimates of blood lead levels can be obtained on repeated occasions, even starting with maternal PbB levels prior to birth, as well as PbB levels in the umbilical cord at birth. PbB levels can then be estimated for example every six months, to help keep track of changing levels of exposure, and as the child develops, its mental performance can also be assessed. A study of this type is longitudinal in the sense that it monitors developments through time, and is prospective in the sense that the child is scrutinized before any adverse symptoms appear, to try to establish whether elevated PbB levels precede IQ deficits or whether the opposite pattern emerges.

One of the benefits of collecting data in a time-series is that one can estimate average blood lead levels over a period of time, and thereby provide a more accurate estimate of the long-term cumulative body lead loads. It is both understandable and fortunate that the scientific community increasingly came to recognize the value of such studies and the new information which they provide.

The reliability and relevance of blood lead measurements can itself, however, be a complex issue. If the members of a community are exposed to relatively constant levels of lead, as might occur if for example drinking water provided their dominant source of lead, then it is reasonable to expect that their PbB levels will remain relatively constant, and provide a relatively good indicator of their accumulated body lead loads. If one or more of those conditions is not met, for example if exposure to lead arises from sanding down old paintwork, then their relative PbB levels would be significantly less informative. This goes some way to explaining why, as Chisholm has pointed out, '...children with identical PbB values rarely show identical toxicological responses.'[13]

Another important consideration is the time at which blood samples are collected. Several of the studies on the possible relationship between PbB levels and neuro-psychological performance which were conducted in the 1980s suffered from a serious flaw because some of the blood samples had been collected several years prior to the psychological assessment of the children concerned.[14] It would be unwise to place much weight on the results of those studies. Furthermore, as time has passed, standards of chemical analysis have improved in several respects. Firstly the sensitivity with which measurement can be made has increased, so that ever lower levels of lead contamination can be measured with ever greater accuracy, and secondly, scientists have also learnt a great deal about how easy it can be to contaminate their samples and measuring instruments, and about the detailed precautions that need to be taken if contamination is to be avoided and precision obtained. One consequence of this process of learning is that, for example, an estimate of a specific PbB level obtained in the mid-1990s is likely to be far more reliable than a similar estimate made 10 or 20 years earlier.[15]

Units

Those reservations apart, PbB measurements are almost always the best available indicator that we have for lead exposure. There are several sets of units in terms of which blood lead levels are described and reported. Scientists sometimes talk about lead constituting so many parts per million (ppm) of blood, by which they mean parts by weight rather than by volume. The most commonly used unit, however, is the number of micrograms (µg) of lead in a tenth of a litre of blood, which, as explained above, is abbreviated to µg/dL. For the remainder of this discussion, those are the units, and that is the abbreviation, which will be used.

Collecting blood samples

Useful though blood lead measurements may be, it is important to appreciate that some methods for collecting blood samples can be more reliable than others. The easiest way to obtain a small sample of blood involves simply pricking a finger with a needle, and collecting the relatively small volume of fluid which emerges. A more difficult procedure involves puncturing a vein, and drawing into a syringe a greater volume of blood. It is possible to contaminate samples when using either method, or indeed with any method, but the scope for contaminating a finger prick test is far greater than when puncturing the vein. The surface of the finger might be contaminated with lead, thereby adding to the apparent level of lead in a resultant sample; on the other hand the clear fluid (which isn't blood) that can be found under the skin can dilute a blood sample to produce an underestimate. Scientists are therefore right to treat a blood sample obtained from a vein (a so-called venal sample) as more reliable than a finger prick, but understandably collecting venal samples provokes far greater resistance from potential donors, especially young children.

It is hard enough to collect a venal blood sample, but it is even harder to persuade an adult, or even more so a child, to provide a regular series of blood samples; however when they do comply, the results can be especially informative. It has really only been during the early 1990s that scientists have been able to discover how useful a sequential set of PbB measurements can be when compared to one-off measurements. When long-term prospective studies are conducted, and blood lead measurements taken at, for example, six-month intervals for six years, it is quite straight-forward to calculate what have come to be known as 'lifetime average' blood lead levels.[16] Since the late 1970s, the scientists who have tried to defend the lead industry have argued that lead should not be indicted as a cause of poor mental performance because the available evidence only showed that there was a correlation between elevated body lead loads and poor cognitive performance.[17] To establish that it was the elevated lead load which was causing the poor performance, it would be necessary first to show that the elevation of the lead load preceded the poor performance. The scientific community rose to that challenge and collectively designed long-term forward-looking studies, raised the relatively large amounts of money that would be required, and then conducted, analysed and reported on them. By the late 1980s, the irony had become clear. A tactic designed to weaken the case against lead had backfired and strengthened that case. It is now possible to estimate life-time average blood lead levels as well as taking single instantaneous measurements, and the evidence which emerged when the former approach was introduced strengthened the case that had already been established by using just the latter.

Use of tooth lead measurements

Several groups of scientists have argued that the measurement of lead levels in teeth can provide a useful index of long-term lead exposure, especially when compared with single blood lead measurements.[18] For obvious reasons, tooth lead measurements are to 'PbT'. Despite the fact that PbT levels can be used as an indicator of the body's lead load, they suffer nonetheless from several important limitations. It is, in principle, possible to take a blood sample at any time, but teeth samples only become available when they are shed naturally, or when subject to dental extraction. Teeth which are deliberately extracted by dentists are typically damaged in some important respect, and therefore cannot be relied upon to provide an accurate indicator of a body's lead load. The only teeth which have been deemed capable of providing a potentially relevant indicator are undamaged milk teeth from children who shed them spontaneously as their adult teeth push through. These teeth, however, become available only when they are ready, often at a time which is not particularly useful to the scientists trying to study those particular children.[19] Milk teeth are normally shed between the ages of five and eight, while scientists might want to study them at other stages in their development.

A group of British investigators reported, moreover, in 1983 that, in the population which they had studied, children who had not donated teeth were more likely to show behavioural disturbances and to have lower cognitive test scores than were those who donated their teeth.[20] The implication of this finding is not only that a sample of donors might be unrepresentative of the broader population, but that it may be systematically biased. The direction of that bias suggests that if we were to rely solely on PbT measurements using voluntarily donated shed teeth there is a risk that we might underestimate the toxicity of lead.

The story is slightly more complicated because the lead which accumulates within teeth is not uniformly distributed; different concentrations of lead can be found in different parts of a tooth. Some researchers have separated a layer of dentine from the rest of the teeth, and have analysed that material for lead, while others have sought to estimate the average levels of lead in entire teeth. There are, moreover, systematic differences between upper and lower jaws, and between different types of teeth such as incisors and molars.[21] This entails that, while tooth lead measurements may possess some desirable features, particularly when compared with single blood lead measurements, considerable care needs to be taken to obtain accurate and reliable results.

Some commentators seem to think that there might be a relatively simple relationship between PbB levels and PbT measurements, but far too little is known about the relationship between these two variables to suppose that one could reliably convert from one to the other.[22]

Use of hair lead measurements

One possible alternative to both blood and tooth samples might be the measurement of the levels of lead in hair. Hair has the considerable advantage of being readily available and painlessly accessible, and once collected the samples remain stable for long periods of time. The main drawbacks with estimating the levels of lead in hair are that it is all too easy to contaminate the hair which is sampled, and very hard to distinguish the level of lead intrinsically present in a sample of hair from contamination which adheres to its surface. This latter problem is so severe that none of the important studies on the effects of lead on health have attempted to rely on hair lead levels as an indicator of body lead loads.

Recently, however, a team at Surrey University have developed a protocol which carefully prescribes from where on the head a hair sample should be taken, and how it should be collected, stored and analysed in order to avoid contamination.[23] This may be a promising line to investigate, but evidence of the toxicity of lead at low levels has emerged from studies using PbB and PbT measurements, and little reliance has yet been placed on measurements of levels of lead in hair.

Debates about lead toxicity

When the US and the UK are compared, it is evident that the debates about the toxicity of lead since the 1970s have shown quite different patterns. The debate in the US has been rather more straightforward than that in Britain. To a first approximation, the former has been a three-cornered argument. On one side, there has been a group of independent scientists some of whom have worked relatively closely with anti-lead campaigners. The lead industry and its representative trade association have formed a second apex of the triangle, with the US Federal government completing the trio.

The lead industry in America has two main representative organizations. One, the Lead Industries Association or LIA, seeks to represent the commercial interests of the industry directly, and across the entire range of industry-related issues. A separate, and more scientific, organization is the International Lead Zinc Research Organisation usually referred to by its acronym ILZRO. ILZRO monitors the scientific debate, and participates in it, partly by commissioning research which it deems useful to the lead industry.

In the mid-1960s the US government adopted a position approximately midway between those who argued that the amounts of lead then reaching the US public were dangerous and the lead industry, which insisted that it was innocuous. As time has passed, however, and the evidence of lead's toxicity has strengthened, deepened, and become more robust, the position of the US government has progressively shifted. That shift has, in large part, been constructed on the basis of a strengthening consensus amongst American scientists about the toxicity of lead at low levels of exposure. Those shifts have been reflected in successive changes in the so-called 'level of concern' for PbB levels which are set by the Centre for Disease Control (CDC). This 'level of concern' is intended to provide a benchmark by reference to which judgements can be made as to whether or not, in this case, a child should be counted as suffering with lead poisoning. The evolution of the CDC's level of concern for young children is illustrated in Figure 2.2.

Figure 2.2 can be interpreted in either, or both, of two ways. One is that the US government has progressively shifted its position by becoming increasingly close to the anti-lead groups; the other is that they have acknowledged the clear implications of developing scientific knowledge. The US government and the anti-lead campaigning organizations do not yet, however, see entirely eye to eye. It is not altogether uncommon to find the

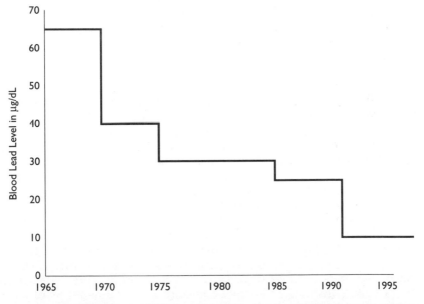

Source: Centre for Disease Control (1991) *Preventing Lead Poisoning in Young Children: A Statement by the Centre for Disease Control*, CDC, Atlanta

Figure 2.2: *Childhood blood lead levels considered elevated by the Centre for Disease Control and the Public Health Service, 1965 to 1996*

anti-lead lobby criticizing the government for not being sufficiently strict with the lead industry, while the industrial lobby argues that US government regulatory agencies are unduly and excessively draconian.[24]

The pattern in the UK on the other hand differs from that in the US in two key respects. Firstly, despite the considerable developments in toxicological science over the past 25 years, the UK government has hardly shifted its position. In 1981, six years after the US CDC lowered their level of concern to 30 µg/dL, the British government, in the form of the Department of Health, advised that any child with a blood lead concentration over 30 µg/dL should be followed-up.[25] A further modification occurred in September 1982, when the Department of the Environment (in collaboration with the Welsh Office) sent an advisory circular to local authorities in England and Wales.[26] The British government then advised that if a child exhibits a PbB level above 25 µg/dL, then an investigation should be conducted to identify and then reduce the child's exposure. When that change occurred, in 1982, for a short time the UK's action limit was slightly more stringent than that in the US. That was an anomalous state of affairs, but it should not be interpreted as indicating that the British government were taking the problem more seriously than their American counterparts. An explanation of how that anomaly arose is provided in Chapter 5. Despite the subsequent accumulation of overwhelming evidence that adverse effects occur in young children at lower levels, even down to 10 µg/dL, the British government has continually failed to respond, and there is no evidence that the British are even contemplating a revision.[27]

The views of the UK government on the lead issue have been, and remain, remarkably close to those of the British lead industry, especially as represented by the London-based Lead Development Association or LDA. The British government has occasionally paid lip-service to the idea that care needs to be taken to protect public health from lead, but their actions have never matched their rhetoric. Nonetheless the British government likes to assert that it has been taking due care; this book documents the falsity of that claim. In the UK, there has been no effective campaigning group addressing lead pollution in a comprehensive fashion since the tax differential on unleaded petrol was introduced by Nigel Lawson when Chancellor of the Exchequer in 1987. Prior to that date, CLEAR, the Campaign for Lead-Free Air, was very active and effective. CLEAR initially raised the political price of the government's stubbornness, during the period when the British government resisted the introduction of lead-free petrol. CLEAR then created the conditions in which the government had a political incentive to shift taxation policy so as to create the illusion that it was serious about environmental issues. Since then, however, CLEAR has gone out of existence, and only Friends of the Earth and Foresight (a rather smaller group concerned with preconceptual care) have campaigned publicly for reductions in exposure to lead. The British government's policy on lead is, however, becoming increasingly untenable.

It has not been too difficult to identify the strategy which the lead industry has been following both in the US and in the UK. It is, at any rate, not too hard to keep track of its public utterances, but it is far harder to know what influence industry has been attempting to exert behind the scenes. In the US, in 1991, ILZRO explained that it was trying to point '...out the inconsistencies and uncertainties in the lead effects literature...' so as to '...suggest that the impact of low level lead exposure *may not* have the adverse health impact claimed by the EPA.'[28] (emphasis added). ILZRO adopted that tactic, but without any conspicuous success. In Britain, there has not appeared to be any need for the LDA to try to persuade the Department of the Environment to neglect the topic, since the DoE has cheerfully neglected the issue unprompted. Furthermore, the civil servant primarily responsible for lead policy at the DoE recently left the Department and accepted a job with the LDA's close associate, the International Cadmium Association, which shares offices with the LDA. There might be a case for the Nolan Committee on Standards in Public Life to review the policy which allows a senior civil servant to move so readily from responsibilities for public policy to employment in the industry which he so recently helped to regulate.

ILZRO's approach to the lead toxicity debate has been relatively modest, but there is no evidence which indicates that lead is safe at low levels of exposure, so all ILZRO can realistically attempt is to highlight and emphasize the limitations in our current knowledge. Their approach relies primarily on drawing attention to the inconsistencies and uncertainties in the data and the results of many studies. ILZRO argued, for example, in 1991 that 'If lead were having a strong adverse effect, the effect should show up much more clearly and be consistent across ... similar studies. Since that is not the case, it can only be concluded that if there is an effect of low level lead exposure on neurobehavioral development, the effect is minimal.'[29] In 1991, that line of argument had some plausibility, but in the interim it has diminished. Over the last six years, as the results of the five main prospective studies have accumulated, a far more consistent pattern of evidence has emerged, and the lead industry has shifted its tactics. Rather than trying to argue that low level lead exposure is safe, they have focused their energies on arguing that they are not responsible for the sources of exposure. Since most children with elevated blood lead levels suffer contamination via leaded water pipes and from old leaded paint, the industry is trying as far as possible to dissociate itself from any responsibility for dealing with the residual problems occasioned by the actions of its predecessors.

ILZRO have a point, nonetheless, in that not all studies of the effects of lead on human health have been equally well designed, conducted and interpreted. If, at any point, therefore we want to assess the strength of the evidence for, and against, the toxicity of lead we should not just follow the relatively simplistic approach of counting the number of studies which indicated a hazard from lead at some particular level of exposure, and compare that with the number of studies which did not indicate a hazard at the same level of exposure. Judgements concerning the relative significance of different and competing studies need to be based rather on an assessment of their strengths and limitations. If it turned out that all the conflicting studies were of equal power and all potentially able to detect an adverse effect, then the fact that they were in conflict might undermine the case against lead. If, on the other hand, there are significant differences between the relative power and sensitivity of the competing studies, we may be able to determine the extent to which the levels of lead, to which some children are currently exposed, are dangerous.

Not all the scientists who have investigated the putative adverse effects of lead on human health have been aligned with the critics of the lead industry. Some investigators have worked closely with the lead industry, and some have chosen to accept research funding from the industry, and to argue their case in a variety of settings. One of the most active members of that group has been Clare Ernhart, and she argued in 1992 that 'The few studies with positive findings are not necessarily those with highest lead levels, the greatest ranges of exposure, the largest numbers of subjects, or the most careful designs.'[30] This is a particularly forceful point, and if it were correct then low levels of lead may deserve to be exonerated, but since I shall show in Chapter 3 that Ernhart's claim is mistaken, a very different conclusion is unavoidable.

In short the debate has been one in which the lead industry and its supporters have progressively had the toxicological ground cut away from underneath them, yet they continue opportunistically to shift their ground from toxicological considerations to the levels of exposure in the population and further on to an argument about who should pay to deal with old uses of lead. There are, however, two tactics which they have never entirely abandoned. One has been to argue that there is a safe threshold below which lead exposure is entirely innocuous, and the other that lead may even be essential for human health.

Is there a safe threshold for lead?

Considerable political importance attaches to the debate about whether or not there is a threshold of exposure below which lead becomes toxicologically insignificant and/or socially acceptable. Policy-makers concerned with the protection of public health often ask whether or not there is some threshold level of lead in the body below which significant adverse effects do not occur. If scientists were able to specify a level of, for example, lead in the blood below which exposure might be deemed safe, then policy-making would be noticeably more straightforward than it is. In practice, the scientific debate is slightly more complicated because the question concerning the possible existence of a threshold can be posed in relation to each of the different parts of the body which lead has been found to damage. There is, for example, evidence of a threshold below which adverse effects on blood pressure seem to cease, but that issue is separate from the question of whether or not there might be a threshold for effects on the kidney or the mental function of young children. If thresholds can be found for a range of different adverse effects, the one by reference to which policy should be judged is the threshold for the adverse effect which occurs at the lowest level of exposure; and where thresholds are not found, policy should be evaluated by reference to the effects found to occur at the lowest levels of exposure.

In Britain there has been hardly any discussion on the topic of whether or not a threshold exists or where it might be found. In the US and elsewhere, however, the topic has been earnestly debated. At a World Health Organisation meeting in Melbourne, Australia in 1992 a gathering of some of the world's foremost experts concluded that 'There is insufficient evidence from the published data to draw any conclusion about the possible existence or otherwise of a threshold...within the range of exposure studied (blood lead 0–40 µg/dL).'[31] The US government's Agency for Toxic Substances and Disease Register (ATSDR), referring to the same evidence, placed the emphasis slightly differently and asserted that there is no known safe level of lead.[32] By contrast, ILZRO has argued that the conclusion that there is no safe threshold for exposure to lead in humans

'...strains credibility, since lead is a natural element, has been used commercially for thousands of years, and is found in the bodies of even the most remote populations at levels close to the mean levels in the US today.'[33]

None of these three considerations, however, is convincing or even relevant. The fact that lead is an element is entirely irrelevant to the question of whether or not it is toxic. Arsenic and radon are naturally occurring elements but no less toxic for that. If anything, the elemental nature of lead counts against lead rather than for it. Because lead is an element, it is indestructible and cannot be detoxified by any process of chemical or biological degradation. The fact that lead has been in commercial use for thousands of years, and that it is present in human bodies throughout the world, does not provide evidence either of safety or for the existence of a toxicological threshold.

If the reason why no evidence of any threshold has emerged was that scientists had not looked thoroughly, then no reliance could be placed on the absence of that evidence. It is never scientifically legitimate to interpret the mere absence of evidence as if it provided evidence of absence. Considerable weight, however, must attach to the fact that scientists have searched avidly for a threshold but failed to find one. As the science improves, evidence of a threshold may eventually emerge. In the interim, however, the lead industry has argued that even if we don't know where a threshold is, there must be one because some minimal level of lead may play an essential role in human biochemistry.[34]

Essentiality

The claim that human bodies have no need whatsoever for lead is frequently encountered. Goyer, for example, states '...there is no demonstrated biologic need...' for lead.[35] Similarly an official committee of British experts has remarked that 'Lead and its compounds ... has no known physiological function...'[36] The suggestion that lead might, nonetheless, be essential has been advanced, for example, by ILZRO. The evidence which ILZRO cites, however, derives only from laboratory studies using animals such as rats, mice and pigs. There is no corroborating human evidence.[37]

Schwartz fed a synthetic diet to rats both with and without lead supplementation and concluded that adult rats fed a lead-supplemented diet had a higher rate of weight gain than similar animals on a low lead diet.[38] Schwartz tentatively concluded that lead may be required, by adult rats, at about 1 ppm in the diet. The implications for human beings, and particularly infants, remain obscure. Schwartz's results are hard to interpret because too little information has been provided about how the animals' diets were manufactured. The synthetic diets were designed to have very low levels of lead, but we do not know how the very low lead levels in the synthetic diets were achieved. This is important because it is hard to envisage any chemical process which could reduce lead levels without also simultaneously reducing the levels of several essential minerals such as zinc, iron and manganese. A severe reduction in the intakes of those essential minerals could by itself account for poor animal health.

In a few animal studies, evidence has emerged which some commentators have interpreted as showing what they call 'homeostatic storage'. What this means is that when lead levels in the blood fall to very low levels, lead stored in other tissues is mobilized into the circulating blood.[39] A similar pattern has been found, for example, with some essential nutrients and minerals. The fact that lead may migrate from sites of long-term storage into the blood does not, by itself however, indicate essentiality. The fact that the tissue distribution may tend toward some equilibrium does not demonstrate that equilibrium level to be optimal. Not all biochemical changes are adaptive, and not all adaptations are ones which enhance viability. The fact that lead can migrate from long-term storage, for example from bone tissue, into the blood tells us something about physiology and biochemistry. The suggestion which ILZRO is making, however, that human beings are healthier with some lead circulating in their blood than they would be with no lead whatever, is entirely unsupported by the evidence, and ILZRO is misconstruing the evidence in a self-interested and unjustifiable manner.

Even if scientists were eventually to conclude that they had identified a level of lead below which it would be imprudent to fall, no-one in the US or the UK appears to be in danger of suffering from a lead deficiency. The entire argument about essentiality therefore deserves to be interpreted as a rhetorical evasion rather than as a serious contribution to policy-relevant science.

The next chapter explains how, over the last 25 years, the evidence that lead is seriously neurotoxic at current levels of exposure emerged, accumulated and strengthened, and why therefore a consensus has grown within the scientific community that action is required to reduce levels of exposure.

3

The Neurotoxicity of Lead

One of America's leading public health experts, Joel Schwartz, has argued convincingly that 'More is known about the health effects of lead than about those of any other environmental pollutant.'[1] The discussion in this chapter is intended to outline the main highlights of the evidence concerning the toxicity of lead, but there is no intention to be comprehensive. The chapter will focus on those parts of the toxicological debate which relate to the most critical effects at the lowest levels of exposure in the most vulnerable sections of the population.

This means that there will be no space for any discussion of the adverse effects that lead can exert on numerous bodily systems. There are, for example, levels of exposure at which lead is known to damage the synthesis of haemoglobin which in turn can cause a cascade of consequent problems, for example in the liver.[2] It is also known to damage neurological functions, to cause kidney damage, to increase blood pressure, to damage reproduction and to interfere with innumerable enzymes.[3] High levels of lead exposure are known to be fatal. Many people have not survived having their blood lead levels elevated to 130 µg/dL.[4]

Since public policy on the control of lead pollution should be designed to ensure that the most vulnerable groups are protected from the effects of both chronic and acute exposure, the remainder of this chapter will concentrate on just one aspect of the toxicity of lead, namely the possibility of neurological damage in young children. Since bottle-fed babies receive almost 90 per cent of the weight of their diet in the form of tap water, such infants living in homes served by leaded plumbing are exceptionally vulnerable.[5]

Lead and neurotoxicity: the crucial questions

There is general agreement that, to the best of our knowledge, the most sensitive effects from lead occur in infants and children, and they involve the nervous system.[6] As long ago as 1914 Thomas and Blackfan showed conclusively that lead is a neurotoxin and that lead poisoning could cause encephalopathy (ie brain damage) with symptoms including convulsions and coma followed by death.[7]

Until the 1970s it was, however, generally assumed that increased lead levels were of little clinical importance if there were no overt signs of poisoning, and if PbB levels were below 50 or 60 µg/dL.[8] During the last 25 years or so, a heated debate has ensued over the existence and severity of adverse neurological effects of lead, primarily in infants and children with blood lead levels below 50 µg/dL.

The crucial questions are: how, in general, might one try to detect and to estimate neurotoxic effects of lead at lower levels of exposure, which effects should investigators be looking for, and how can potentially relevant neurological indicators be identified and selected? Given that, at high levels, lead causes fatal brain damage, low levels of lead might be expected to cause relatively milder brain damage. Since we cannot inspect the condition of human brains directly, it is important to try to decide which overt changes, which may be cognitive or neurobehavioural, could most effectively indicate subtle forms of brain damage. Scientists use the phrase 'cognitive indicators' to refer to those related to mental and intellectual functioning, while the expression 'neurobehavioural indicators' is used to refer to physical activities rather than those which are primarily mental.

In advance of conducting programmes of research it was impossible to tell which type or types of indicators should be selected, given that the objective has been to identify the most sensitive and reliable indicator of the damage which low levels of lead might provoke. In practice, as Smith et al have argued, 'The test batteries employed in lead studies have tended to be large and comprehensive, and have been described as following *the blunderbus approach; that is, if you don't really know what you are looking for, try to cover everything.*'[9] (emphasis added).

Similarly Yule and Rutter explain that 'In the absence of any agreed plausible model of neuropsychological processes affected by low level lead exposure, most investigators have resorted to a *pot-pourri* mixture of broad measures of general intelligence together with an assortment of less well validated measures supposedly sensitive to neurological dysfunction.'[10] These two remarks are broadly correct. When research commenced, scientists were unable to predict in advance precisely how the possible neurotoxicity of low levels of lead might be exhibited and consequently they tried to use a wide variety of potential indicators, but in due course investigators increasingly focused on some relatively narrow aspects of intellectual performance.

One problem which confronted, and to some extent continues to confront, research scientists is rather tricky because human intellectual performance is known to be influenced by a very wide range of social, psychological and biological factors, not all of which have been fully identified or characterized. It would be a bold scientist who pretended to be able to explain fully all the factors that determine, or even influence, the intellectual performance of individuals or groups. Consequently, an attempt to detect the effects which marginally elevated levels of lead in the body might be having is rather like trying to see whether or not there is a signal which can be detected against the background of a considerable volume of noise.

In practice, when researchers have sought a suitable indicator of mental and intellectual performance, they have chosen one or more of a variety of test systems which are supposed to estimate what experimental psychologists refer to as an intelligence quotient or IQ. There are several different types of IQ tests, not all of which are equally reliable or suitably validated.[11]

The debate about the meaning, utility and relevance of IQ tests is almost as old as the tests themselves. Some scientists take the relatively simplistic view that intelligence should be defined as that which IQ tests measure. That position is simplistic, in part, because it fails to provide any basis from which to explain why any one IQ test system should be preferred over any other, a question which is unavoidable because competing test systems can yield entirely different results.

Disputes about the meaning of IQ scores

In America, Bellinger is able to say 'Everyone thinks they know what IQ is and that higher is better.'[12] British attitudes to IQ tests and scores are far more cautious and hesitant. Furthermore, in America economists think that they can assign dollar values to IQ change.[13] They do that firstly by showing that children with high IQ scores when at school earn, on average, higher incomes after leaving school. They then calculate by how much incomes vary with scores achieved in IQ tests while at school. That approach is largely discredited in Britain, and is thought of as being too crude; but then Britain is altogether less meritocratic than the US.

The protagonists in the debate concerning the putative neurotoxicity of lead have typically assumed that it is possible to disentangle, for example, genetic from environmental factors influencing intelligence and mental performance, and moreover that it is possible to distinguish between the relative contributions of different environmental factors. In making that assumption, scientists have often ignored a complex but important debate concerning the very possibility of independently characterizing and then separating such factors.

There has, in point of fact, been an enormously heated debate, separate from the issue of lead safety, as to whether or not it is possible to distinguish between, or even to estimate the relative contributions of, such genetic and environmental factors as may influence performance in IQ tests.[14] Within the toxicology community there has simultaneously been a debate about the levels in the blood at which lead may be neurotoxic, and whether low levels of lead are exerting an adverse effect on children's performance in IQ tests. While those two debates are relevant to each other, they have failed to intersect.

Scientists who conduct studies on the effects of lead on mental performance have identified innumerable factors which they suspect influence performance in those IQ tests. It is therefore difficult to design, conduct and interpret studies which are sufficiently reliable and thorough that they can provide conclusive evidence that lead at low levels, *and lead alone*, accounts for some observed variance within the results of the IQ tests. Evidence of such effects can be found, but finding it requires using complex statistical methods to analyse the data obtained.

Even amongst those who assume that IQ measures can be meaningful, and that environmental and genetic factors can be disentangled, there are disagreements about the relative utility of different systems and the limits within which they can be used. According to Fulton, for example, 'It is only from about 6 years and upwards that tests have satisfactory reliability and predictive value. Yet the brain is probably most vulnerable to damage in the first two years of life when testing is less satisfactory.'[15] Similarly, a working party of the UK's Medical Research Council (MRC) stated in 1988 that the Bayle Mental Development Index (MDI) scale (which has been extensively used in studies of the putative neurotoxicity of lead) was '...not originally designed to be sensitive to small and subtle differences in performance between children.'[16] If that is correct it makes it extremely difficult to detect effects in precisely the context in which they may be most significant. Important questions therefore arise concerning the relative merits of competing methods for assessing mental development and performance, and the differing stages and ages at which they might be appropriate.

Towards the end of 1992, a meeting of international experts was held in Melbourne, Australia as part of the preparation for an evaluation of the toxicity of lead being conducted by the World Health Organisation. The participants agreed that '...intelligence tests do not provide perfectly reliable scores. Measurement error is always present; and tests do not measure all aspects of intelligence...but they are the best available measure of broad intellectual functioning. Furthermore, the limitations outlined above are more likely to bias outcome measurements towards a null effect.'[17] That last sentence is important because it implies that a lead-related deficit in mental performance will be difficult to detect using IQ tests, and therefore evidence of such effects should be interpreted as an underestimate of its true extent and significance.

The meaning of IQ scores

Whatever it is that IQ tests measure, it is not a natural constant like the speed of light or the refractive index of water. It should be thought of rather as a social construct. IQ tests have been designed and refined so as to produce a population average result of 100, with a standard deviation of 15.[18] IQ scores have that pattern not because that is a fact of nature, but because the tests have been engineered to produce that result.

IQ scores are certainly what mathematicians call 'ordinal' in the sense that they generate a set of integral numbers, but that does not mean that IQ scales are what mathematicians refer to as 'homogeneous', 'uniform' or 'metric'. Results on an IQ scale would be homogeneous if, for example, a five-point difference had the same meaning at all points on the scale. A difference between 80 and 85 is not, however, necessarily the same as the difference between 120 and 125. Similarly, IQ scores are not necessarily 'metric' in the sense that a person scoring 120 is one third more intelligent than someone scoring 90. Furthermore, if one person scores 150, a second scores 120, and a third scores 90, we cannot conclude that the difference between the first and the second is the same as the difference between the second and the third, nor that the difference between the first and the third is twice the difference between either of those two and the one in between.

Steven Rose is one scientist among many who has argued that all IQ test systems inevitably presuppose a set of cultural and social factors which are unrelated to an individual's mental development. Rose has pointed out that

> *It is possible to devise tests on which the working class score higher than the middle class, or blacks higher than whites. However such tests are discarded, and the fact of the black's better score is actually used, for instance, by Jensen, to claim that it represents a lower-order intellectual skill...* [19]

Similarly, Ryan has concluded that '...IQ is not, and could not be, a measure of cognitive abilities abstracted from all social and motivational factors. In as much as IQ tests measure anything, they measure the likelihood of educational and social success in a particular society.'[20]

It is both possible, and important, to acknowledge the culturally and socially bound nature of IQ tests without entirely dismissing their utility, but it does imply that the results of IQ tests need to be interpreted with considerable care. For example, comparisons between children from very similar social groups are likely to be more significant and reliable than comparisons between children from markedly different social or cultural backgrounds. Comparisons made at two different times for a given group are similarly more significant than comparisons between different groups at any one time.

Confounding factors

A further set of difficulties arise because there are many other social, psychological and biochemical factors which can affect mental functions, and which can therefore complicate any study of the putative relationship between lead-loads and IQ scores. It is important to try to identify, both in general terms and in each particular study, what the confounding factors might be. To a first approximation, there are two approaches to the identification of confounding factors. There can be an attempt, in advance of conducting a study and generating the data, to specify those factors which are believed to be candidate confounders. Alternatively researchers might simply try to measure a very wide range of variables, and then try, in a *post hoc* fashion, to analyse the data to determine which of those factors actually were exerting an influence.

Given that performance on IQ tests, and similar procedures, is a function of a complex and heterogeneous set of factors, those factors which confound the relationship between performance and indicators of body lead load need to be both properly identified and handled. There are two key dangers which researchers need to try to avoid, even after putative confounders have been identified. If due weight is given to a confounding factor then it will have been properly handled, but if too large a proportion of the variance in the dependent variable is attributed to a confounder, then it will have been over-controlled, while the opposite mistake is termed under-controlling.

Methods of data analysis

Studies of the effects of lead on mental development and performance typically generate relatively large quantities of data concerning not just levels of lead in blood or teeth and IQ measurements but also a complex variety of potentially confounding factors. The question of how these data can, and should, be analysed needs to be addressed.

As Yule and Rutter have pointed out, if a large number of psychological measurements are taken, and the evidence indicates that some variables are apparently affected while others are not, the problem arises as to which method or methods of statistical analysis would be appropriate. The two main alternatives are referred to by statisticians as 'univariate' and 'multivariate' methods of analysis. Univariate analysis is the simpler of the two approaches because it examines the way in which one factor influences another, whereas multivariate analyses explore the ways in which a multitude of factors might influence a particular outcome. Since neurological abilities and mental performances can depend on many variables other than just exposure to lead, it is preferable for multivariate statistical methods to be used when trying to determine whether or not lead is exerting an adverse effect.[21]

As Pocock and Ashby have explained

> *A child's IQ is influenced by such factors as the degree of parental interest in the child or the stability of marital relationships... It is impossible to quantify all non-lead influences on IQ so that adjustment for confounders can never be complete. Conversely, one has to be careful not to over adjust by using covariates which may reflect*

lead exposure. For instance, it would be inappropriate to use pica *(ie, ingestion of non-food items) as a confounder since this may be directly responsible for an increased body lead burden.*[22]

Protagonists in the debate about the neurotoxicity of lead often write in terms of 'controlling for' or 'correcting for' factors supposed to be confounders. Since it is extremely difficult to tell whether or not the confounding factors have been properly identified and controlled for, the use of those two expressions carries with it the risk of conveying a misleading impression that a greater degree of precision has been achieved than can legitimately be claimed. Consequently, the more neutral expression of 'adjusting' for putative confounders is preferred in this text.

Statistical power and sample size

As with all epidemiological studies of putative chronic problems, some of the key statistical issues arise in relation to the design and conduct of the study rather than the subsequent analysis of the resultant data. The key factor which determines the statistical power of any study to detect an effect is the size of the sample being studied. The US Environmental Protection Agency (EPA) has argued, in particular, that to obtain statistically significant evidence sufficient to indicate the detection of a modest effect of lead on IQ, it would be necessary to study a population consisting of at least 400 subjects.[23] Very few of the studies so far conducted have had sample sizes remotely approaching that figure, and consequently many studies may not have possessed sufficient statistical power to detect an effect were it to be occurring. As a result, some of those which failed to detect an effect should perhaps be categorized as inconclusive rather than as genuinely negative. An interesting corollary follows however, namely that if and when it becomes possible to pool the results from several studies, and to treat the sum of the sizes of their samples as if it were one big sample, then the statistical sensitivity of the analysis can be correspondingly increased. For reasons which should become clear before the end of this chapter, this latter approach has paid substantial dividends.

Statistical analyses

Any properly designed study, in this context, should try to test a hypothesis. Typically that hypothesis will be that elevated lead loads can have an adverse effect on neurological functioning. The statistical analysis of the resultant data will be intended to distinguish between that hypothesis and what is termed the 'null hypothesis', namely that lead has no observable adverse effect under the conditions of study. The purpose of a test of statistical significance on the data from a lead–IQ study is to assess the extent to which the observed data are inconsistent with the null hypothesis. Each significance test produces a value for the statistical variable represented by the letter P, where P is the probability of getting the observed association (or something even stronger) in a random sample of children if the null hypothesis of no association were correct. The smaller the value of P, the stronger the evidence for a genuine association between lead exposure and outcome. Professional statisticians have adopted the convention that a value of P less than 0.05 is used to refer to a link which they deem 'statistically significant'.[24] The meaning of that figure of $P < 0.05$, in this context, is that even if lead was unrelated to performance on neuropsychological tests, one test out of every 20 such tests would randomly produce an apparently significant result. That means that there would be a 5 per cent risk of falsely implying a link when one did not exist.

Research strategies in lead neurotoxicology

A wide variety of different approaches have been adopted to investigate the possible adverse neuropsychological effects of low level exposure to lead in children. As Yule and Rutter have explained in their lucid account, to a first approximation they can be separated into six different approaches. These are:

1. clinical studies of children with high lead levels;
2. studies of mentally retarded or behaviourally deviant children;
3. chelation studies;
4. smelter studies;
5. general population cross-sectional studies; and
6. general population prospective/longitudinal studies.[25]

Of those, only chelation studies involve some direct intervention, the other five approaches being observational studies. This classification represents, amongst other things, a broad temporal sequence because, although there have been some periods of overlap, clinical studies of children with high lead levels provided some of the earliest approaches, while for example studies on groups living close to smelters came at a later date, and prospective studies of general population groups represent the most up-to-date approach. The temporal sequence also represents a progressive improvement in experimental power because, other things (such as sample size) being equal, the more recent approaches have possessed greater sensitivity than those utilized at an earlier date. For all of these reasons, these various methodological approaches will be discussed in the sequential order outlined above.

Clinical studies of children with high lead levels

One approach, which was relatively common especially in the early 1970s but which is no longer so popular, was to study children who had been identified in clinical contexts as having elevated blood lead levels. The tactic was to compare some of their neurobehavioural skills with those of a group of otherwise similar children with lower levels of lead in their blood.

Several studies conducted during the 1970s suggested that where lead levels were consistently raised above 60 μg/dL, a decline of three to four IQ points could be detected, even among apparently asymptomatic children.[26] At that stage, however, the evidence of an adverse effect was much less clear in the PbB range of 40–60 μg/dL, but there were some indications of an effect. These early studies were unfortunately very unreliable, partly because they used very small samples, and partly because of a lack of proper statistical control for the confounding effects of physical and social characteristics of the children's backgrounds.[27] By 1980, however, evidence had accumulated to indicate that lead was capable of causing impaired neurological functioning at blood lead levels below those which had traditionally been associated with clinical lead poisoning. During the intervening 16 years, however, research methods have improved to the point where adverse effects are now being demonstrated in children who are not even suspected of being lead-poisoned.

Studies of mentally retarded or behaviourally deviant children

This type of study is similar to the first group, but it starts, as it were, from the other end. A sample of children categorized as being mentally retarded or behaviourally deviant were studied to investigate whether or not their blood lead levels were higher than those in an appropriate control group. These studies consistently found higher levels of lead amongst the dysfunctional children than amongst the controls, but as Yule and Rutter explained

> *All these studies have numerous weaknesses including paucity of data on relevant background variables, arbitrary and uncertain designations of "cause", and missing data. In view of these limitations, no firm conclusions are possible from this research strategy. The consistent finding of higher lead levels in deviant groups [was] provocative but it is uncertain whether they represent a cause, or an effect, or a mixture of the two.*[28]

Therapeutic chelation studies

Chelation is a form of clinical treatment in which a chemical is administered to patients who are carrying unduly high body lead loads. The treatment is intended to extract lead from the body and so diminish the load. The idea of chelation studies is straightforward, namely to test neuropsychological functioning and PbB levels both before and after treatment to see whether a reduction in PbB levels produces any improvement in neuropsychological performance. A study of this type might have the advantage that it could be conducted under relatively strict experimental conditions. If it could then be shown that a reduction in blood lead levels was followed by improvements in neurological function which would not otherwise have been expected, then the result might be important.

One possible advantage of chelation studies is that they enable investigators to address relatively directly the vexed question of the direction of causation; namely is lead poisoning damaging the intellectual performance of young children, or are less bright children simply more prone to absorbing lead from their environment than the more intelligent ones? One of the routes by which babies and young children can absorb lead is by putting in their mouths inappropriate objects such as toys and dirt, which can be contaminated with lead-bearing dust. This practice, which often includes eating dirt, is known as pica. Since chelation studies actually involve lowering PbB levels, and estimating the consequences, they have the potential to show that correlations between elevated PbB levels and lower IQ scores cannot simply be explained away by reference to higher rates of pica amongst the less bright children.

Several precautions would also need to be taken, however.

In particular, it would be desirable to compare the apparent effects of chelation on one group of children with those in a control group which had received a placebo, or inert imitation of the treatment, rather than the authentic chelator. This is important because there is a chance that the children's performance might improve simply as a response to the extra attention which they were receiving rather than as a consequence specifically of the chelation. Unfortunately, none of the published chelation studies have involved that crucial comparison.[29] That is, however, understandable because there are ethical problems about withholding treatment from lead-poisoned children.

Two recent studies, both reported by Ruff and her colleagues in New York, have provided some evidence that lowering body lead loads can have a beneficial effect on performance. The results of one study into the impact of chelation therapy on moderately lead-poisoned children were published in 1993.[30] The investigators recruited 154 previously untreated children aged from one to seven years with PbB levels between 25 and 55 µg/dL. They treated the children with the most commonly used chelating agent, known as Calcium EDTA,[31] and their PbB levels declined markedly. While no impact on IQ scores was detected over a period of a few weeks, once six months had passed, the investigators found a gain of one IQ point for every decrease in PbB of 3 µg/dL, even after controlling for confounding variables.

A similar study published in 1996 of 42 moderately lead-poisoned children around two years of age, found that

> ...the change in standardized score (particularly change in perceptual-motor performance) was strongly related to change in blood lead in children who were iron sufficient at the outset: there was an increase of 1.2 points for every 1 µg/dL decrease in blood lead. There was no such relationship in iron deficient children.[32]

This relationship was valid across the entire range of PbB levels found, from about 40 µg/dL to about half that figure. This suggests that lowering blood lead levels can have a considerable impact on the children's performance in IQ tests, just as long as they are otherwise properly nourished. That condition is important because the chelating treatment can leach out of the body not just lead but other minerals such as zinc, iron and manganese which are not just desirable but essential. Chelation therapy should therefore perhaps best be thought of as having a valuable role to play in treating clinical lead poisoning, rather than primarily as a research strategy in neurotoxicology.

Smelter studies

Children living close to lead smelters provide a particularly suitable group for investigation, especially because the studies can focus on children whose exposure to lead is almost certainly extrinsic to, and independent of, their own behaviour.[33] This is an advantage not possessed by studies of children with clinically elevated blood lead levels or with studies of mentally retarded or deviant children, because in such cases it is difficult to distinguish between causes and effects. Between 1974 and 1985 approximately half a dozen smelter studies were conducted but all of them suffered from methodological problems.[34] The main difficulty was a consequence of the fact that it was hard to ensure that families living near a smelter were not significantly different from potential control groups in some other relevant respects.[35] It is easy to appreciate that the vicinity of a lead smelter or battery factory is not necessarily the most desirable neighbourhood, and the distance from a family's home to the nearest smelter might be an indicator not just of the levels of lead to which they may have been exposed, but of other relevant variables too. Consequently Yule and Rutter explained in 1985 that 'These methodological difficulties severely limit the conclusions that can be drawn from smelter studies; nevertheless ... four of the six studies were consistent with a small (1–5 points) intellectual deficit associated with raised lead levels in the 40–80 µg/dL range.'[36] Despite these methodological difficulties, smelter studies continue to be conducted, and the quality of the research has certainly improved as time has passed. A particularly careful study was conducted in Lavrion in Greece by a team based at Athens University, the results of which emerged in the late 1980s.

The Athens University Medical School lead study
The Greek investigators studied the relationship between performance in IQ tests and blood lead levels in 509 children who lived near a lead smelter in Lavrion.[37] Lavrion is a town of some 10,000 inhabitants. Silver has been mined in Lavrion since approximately 600 BC, and a lead–zinc smelter had been operating there since 1864.[38] The sample of children studied was drawn from four primary schools during the 1984/85 school year. Samples of venal blood were collected and assessments of neuropsychological performance were made, including performance on an IQ scale, and those data were supplemented by information on social and familial factors. The average PbB level in the experimental group was relatively high at 23.7 µg/dL,

and individual estimates ranged from 7.4 µg/dL to the very high figure of 63.9 µg/dL. In their statistical analyses, the researchers examined the relationship between PbB and IQ results using several analytical models which included no fewer than 8, and no more than 23, confounding variables.

The team of investigators concluded that they had detected a consistent decrease in performance on the IQ tests as PbB levels rose, but only above the 25 µg/dL level.

They reported specifically that

> *The full-scale IQ–PbB association was fairly robust using various models with 8 to 23 covariates, resulting in a decrease of 2.4–2.7 units of IQ for every 10 µg/dL increase in PbB...the adjusted full-scale IQ difference between "high" PbB (>45 µg/dL) and "low" PbB (> 15 µg/dL) groups was 9.1 units.*[39]

At that time, that difference was one of the highest ever reported in a study of the possible links between PbB levels and performance on IQ tests.

Interim summary

In summary, this discussion has indicated that these four approaches to investigating the neurotoxicity of lead can provide decisive results only if several conditions are satisfied. Firstly, the population sample must be sufficiently large to permit the detection of a marginal effect, secondly, care needs to be taken to estimate and adjust for confounding factors, and thirdly the issue of the direction of causation needs to be addressed. The first and second issues came to be addressed in cross-sectional studies of the general population, while the third has been dealt with by a subsequent set of prospective or longitudinal general population studies, and it is the results from these two approaches which are reviewed below.

General population cross-sectional studies

It became clear to most researchers during the 1970s that any adequate study of the putative neurotoxicity of low levels of lead would require a carefully controlled investigation of a representative sample of the general population. The resultant studies have been characterized as 'cross-sectional' to distinguish them from a subsequent group which are known as 'prospective' studies. The expression 'cross-sectional' is used in the sense that they attempt to measure body lead loads and neurological functioning at particular points in time, rather than on successive occasions. Prospective studies can have the advantage that, by following the changing fortunes of a group of children, they enable questions of the direction of causality to be addressed which are beyond the scope of cross-sectional investigations of a single occasion. Long-term prospective studies can suffer, however, from another drawback, namely the difficulties in retaining participation by the children and their families.

Between 1972 and 1990 a total of 27 cross-sectional studies of the effects of lead-exposure on children's performance in IQ tests were published.[40] Several of them provided useful evidence indicating that lead, even down to relatively low levels of exposure, can exert adverse effects on performance in neuropsychological tests. Some studies produced equivocal results and some were negative. None of them produced evidence, one way or the other, which was definitive or conclusive. Table 3.1 lists the 27 studies in chronological order.

Little purpose would be served by detailing all of those 27 studies, but it is worth discussing six examples, because they highlight what the scientists were, and were not, able to accomplish at that stage.

Needleman et al's 1979 tooth lead study
A major development occurred with the publication of a study by Needleman and his colleagues in 1979, which used the levels of lead in teeth (PbT) as the main indicator of lead load, and which drew on a large population sample and endeavoured to estimate, and adjust for, a wide variety of potentially confounding factors.[41] This study has subsequently become the focus of an extremely heated and acrimonious debate, the details of which are not entirely relevant in this context. If, in the intervening period, further evidence concerning the neurotoxicity of low levels of lead had not been forthcoming, then the virtues and shortcomings of this particular study would be crucial, but since few commentators or policy-makers now rely on the results of this study as a basis for their current judgements, a final determination on the validity of the results of this study will not be decisive for our purposes.

The investigators obtained shed milk teeth from 70 per cent of a sample of 3329 potentially eligible children based in two of Boston's suburbs. Children were, however, excluded from the investigation if they already had indications of clinical neurological impairment or evidence of overt lead poisoning. The remaining children were classified as to the level of lead in their

Table 3.1 *Lead neurotoxicity studies identified by Needleman and Gatsonis*

Investigator	Year	Number of Subjects C=Controls; E=Exposed	Tissue tested	Lead Effect $P < 0.05$
Kotok	1972	C=25; E=24	Blood	No
Perino and Ernhart	1974	C=50; E=30	Blood	Yes
de la Burde and Choate	1975	C=67; E=70	Blood	Yes
Landrigan et al	1975	C=78; E=46	Blood	Yes
McNeil et al	1975	C=37; E=101	Blood	No
Yamins	1976	80	Blood	Yes
Kotok et al	1977	C=36; E=24	Blood	No
Ratcliffe	1977	C=23; E=24	Blood	No
Rumno et al	1979	C=45; E=45	Blood	Yes
Needleman et al	1979	C=100; E=58	Tooth	Yes
Yule et al	1981	166	Blood	Yes
Winneke et al	1982	C=26; E=26	Tooth	No
McBride et al	1982	108	Blood	No
Smith et al	1983	402	Tooth	No
Winneke et al	1983	115	Tooth	No
Harvey et al	1984	48	Blood	No
Shapiro and Marecek	1984	193	Tooth	Yes
Needleman et al	1985	218	Tooth	Yes
Ernhart et al	1985	80	Blood	No
Schroeder et al	1985	104	Blood	Yes
Hawk et al	1986	75	Blood	Yes
Lansdown et al	1986	C=80; E=80	Blood	No
Hatzakis et al	1987	509	Blood	Yes
Pocock et al	1987	402	Tooth	Yes
Fergusson et al	1987	724	Tooth	No
Fulton et al	1987	501	Blood	Yes
Hansen et al	1987	156	Tooth	Yes

Source: H Needleman and C Gatsonis (1990) 'Low-Level Lead Exposure and the IQ of Children, A Meta-analysis of Modern Studies' *Journal of the American Medical Association* Vol 263, No 5, 2 February, p674

teeth by reference to two samples of dentine, taken either from the same tooth or from duplicate teeth. The purpose of this refinement was to increase the precision of the tooth lead measurements.

Of those in the sample, 58 children were defined as falling in a 'high lead' category (comprising those with teeth containing more than 24 ppm or μg/g of lead) and 100 as being 'low lead' (comprising those with teeth containing less than 6 ppm of lead). About half of the eligible children were not tested, which was a large proportion, and a source of possible bias, but those excluded were reported not to differ with respect to either PbT or teachers' ratings from those remaining in the sample. Both groups of children underwent approximately four hours of detailed neuropsychological testing, which constituted a considerable advance on previous studies. The high lead and low lead groups were broadly similar on background variables but the high lead group was slightly older and slightly more socially disadvantaged than the low lead group.

The authors reported that 'As compared to controls, children with high lead levels appeared particularly less competent in areas of verbal performance and auditory processing...The ability of subjects with high lead levels to sustain attention was clearly impaired, as measured by reaction time at varying intervals of delay.'[42]

As Yule and Rutter pointed out

> *...this was a well-planned, detailed, systematic and well-analyzed study. The lead effect on IQ that was found was small (some 4.5 points after correction) but the fact that it remains after taking into account a range of relevant family variables suggested that the effect represented a true causal influence, and not any form of artefact....inevitably there were a number of limitations to this study but none are sufficient to invalidate the findings...*[43]

The 1979 Needleman study had an enormous impact on the policy debate, especially in the US, but also in the UK and elsewhere. Its impact was partly a consequence of the relatively large size of the sample, and partly because of the care and detail with which the study was conducted and reported. Another important factor, as the above quotation indicates, is that it represented substantial progress in attempts to distinguish systematically between adverse neurological effects of lead, and the effects of the numerous other factors which also influence performance on neuropsychological tests.

This study, and its interpretation, were heavily criticized because it represented such a powerful challenge to the status quo and especially to the lead industry. Kramer, for example, complained that Needleman and his colleagues had failed to show that lead levels preceded lower IQ levels, rather than arising as a consequence.[44] Kramer called for a prospective cohort study comparing two groups of children shown to be neuropsychologically equivalent before lead exposure. There is a sense in which Kramer is correct, namely that one cross-sectional study by itself could not provide a demonstration of causality, in either direction, but it was perhaps unreasonable of Kramer to complain that Needleman et al had failed to demonstrate something which they had not set out to show. The case for prospective longitudinal studies was, and is, a sound one, and in due course the call was met, but that should not be interpreted as implying any inadequacy in the Needleman et al study by the standards of cross-sectional research.

Hall argued that differences in the performance of the high and low lead groups might have been accounted for in terms of marginal social differences, and in a similar vein Cole (of the International Lead Zinc Research Organisation) argued that the low lead children were perhaps more likely to have had pre-schooling, and that variable had not been controlled for.[45] Coplan argued that parental IQ was a potentially confounding variable for which too little adjustment had been made, while Lynam (writing from the Lead Industries Association) argued that Needleman et al did not relate dentine lead to lead exposure.[46] Lynam also argued that the investigators had failed to relate dentine lead to historical blood lead levels adequately. These criticisms drew attention to matters which were not entirely misleading but which in this context were beside the point. Needleman et al had not set out to answer every single possible aspect of the putative neurotoxicity of lead. On the contrary, they set the terms of reference of their study rather carefully, and criticizing them for failing to do what they had not set out to do could perhaps be counted as unfair, especially as no other researchers would have been in a position to meet all of those criticisms either. Needleman's response to these criticisms was, however, dismissive rather than calmly reasoned.[47]

In the intervening years, since the publication of Needleman et al's 1979 paper, the debate concerning the validity of this study has continued and become increasingly acrimonious. This is certainly not the only acrimonious part of the lead toxicity debate, but it is a part which has received most public attention.[48] In response, the Environmental Protection Agency established a committee in 1983 to provide an independent peer review of several contentious studies, and this committee reported to the Director of the Environmental Criteria and Assessment Office of the EPA in November 1983.[49]

The Expert Committee on Pediatric Neurobehavioral Evaluations said of the Needleman et al 1979 paper

> *...that the reported results concerning the effects of lead on IQ and other behavioral neuropsychologic abilities measured for the low-Pb and high-Pb groups must be questioned, due to: (1) errors made in calculations of certain parental IQ scores entered as control variables in analysis of covariance; (2) failure to take age and father's education into account adequately in the analyses of covariance; (3) the failure to employ a reliable strategy for the control of confounding variables; (4) concerns regarding missing data for subjects included in the analyses; and (5) questions about possible bias due to exclusion of large numbers of provisionally eligible subjects from statistical analyses. The Committee conclude[d], therefore, that the study results...neither confirm nor refute the hypothesis that low-level Pb exposure in children leads to neurologic deficits.[50]*

As Palaca has explained

> *...Needleman fought back. He insisted that the ...[expert committee's] conclusions were flawed, and he wrote a spirited, point-by-point refutation of the criticisms levelled at his work. He blasted Grant [the then Director of the EPA's Environmental Criteria and Assessment Office] for printing the report before sending it to him for review, accusing him of violating an agreement he said he and Grant had made. Needleman also performed some new analyses of his original data, and by the time the ...[expert committee's] report was presented to the EPA's advisory panel that would decide on the new lead standards, both Grant and the advisory panel had made a 180-degree turn. Now they were convinced that Needleman's original conclusions were accurate.[51]*

It is now evident that the US EPA, the Centre for Disease Control and the Agency for Toxic Substances and Disease Registry do not share the Committee's initial reservations concerning this study.[52] These matters continue, however, to be hotly contested.[53]

In Pittsburgh on 14 April 1992 an extremely rare event occurred, when the University of Pittsburgh held a public hearing to examine allegations against Professor Needleman by Dr Sandra Scarr and Dr Claire Ernhart.[54] By May of 1992, it was reported, however, that the University of Pittsburgh had rejected the allegations of misconduct against Needleman.[55] A subsequent report indicated that the investigatory panel had been unanimous in its conclusion that no misconduct had taken place.[56]

Writing in *The Journal of NIH Research*, Taylor claimed to have a copy of the panel's report, which he says states that '...Needleman deliberately misrepresented his procedures in the study published in March...1979...But it nevertheless finds him not guilty of scientific misconduct. The board says it found no evidence to support allegations that he improperly distorted his data or methods of analysis to arrive at his conclusions.'[57] According to Taylor, 'Needleman acknowledges that he made errors in describing his procedures in the paper, but he vociferously denies the charge of deliberate misrepresentation.'[58]

In August 1991, an article in *Science* drew attention to the fact that 'Regardless of who is right, the Needleman saga shows how hard it is to put to rest charges from persistent critics, or, conversely, to prove misconduct against an acknowledged leader in a scientific field.'[59] What is clear is that while the work by Needleman et al played a vital role at the time in moving the debate on lead toxicity forward, so much other work has subsequently been published which has resulted in a considerable diminution in the relative significance and importance of that endlessly controversial paper.

The Three London Boroughs Study

In the early 1980s, a team jointly based at the Institute for Child Health and Southampton University studied a sample of six-year-old children in three London boroughs, two of which were relatively suburban while the third was closer to the city centre.[60] Sixty per cent of the children in a sample of 6875 gave at least one tooth. On average, donors of teeth differed from non-donors by being from a higher social class, and having higher scores on cognitive tests, less behavioural disturbance, and were more likely to be female.[61]

When analysing the teeth for lead, the researchers focused their attention on incisors because most of the teeth were incisors, and because the incisors were found to have higher lead levels than other teeth. In an attempt to cope with the known variations between the levels of lead in different types of teeth, the scientists introduced an adjustment to take into account the differences between upper and lower jaws, and between medial and lateral incisors. The suitability of such an adjustment depends on the relative consistency in the variations in lead levels between those various types of teeth, but not enough is known about the consistency of those variations to enable one to be certain whether the adjustments which were made were appropriate, excessive or too slight.

For analytical purposes, the children were divided into three groups, a low-lead group with PbT figures of 2.5 µg/g or less, a medium-lead group with PbT levels between 5 and 5.5 µg/g, and a high-lead group with PbT levels of 8 µg/g or higher. The parents of 97 per cent of the subsample of children agreed to co-operate, and blood lead samples were taken from 92 children. The mean blood lead levels for those three groups were 11.5 µg/dL, 11.9 µg/dL and 14.4 µg/dL respectively, in other words there were far greater differences between the levels of lead in their teeth than there were in their PbB figures. The information which the researchers managed to obtained from both families and children was unusually thorough and extensive by the standards of that time.[62]

Before adjustments were made for confounding factors, there were significant differences in the mean IQ scores of the three groups. The low-lead group scored 108.7, the medium group reached 104.9 while the high-lead group scored 103.7; so for the unadjusted figures, the difference between the high- and low-lead groups was a full five points. After adjustment for a rich set of confounding factors, the investigators reported that their high-lead group had a mean full scale IQ 2.3 points lower than that found in the low-lead group, although the measure of statistical significance of the relationship fell marginally short of the figure (p < 0.05) which is normally deemed to indicate a statistically significant result.[63] Yule and Rutter are very complimentary about the way in which this study was conducted, but less complimentary about the way in which the results were analysed and reported. Briefly, they provide a complex set of reasons for believing that, because of the way in which the covariates were chosen, and because of the large number which were included, '...overcorrection may have occurred.'[64] That is to say, there are reasons for thinking that a more judicious choice of variables in their multivariate analysis would have yielded a more straightforwardly positive result.

The Edinburgh study by Fulton et al
A study based in Edinburgh, Scotland, and conducted in the mid-1980s, was a particularly well-designed and well-executed

project.[65] Edinburgh was chosen because of the acidity of its water supply, the high proportion of homes which received their drinking water through lead plumbing and consequently high levels of lead in the drinking water.[66] The researchers investigated a sample of 855 boys and girls, and gave a detailed examination to a relatively large subsample of 501 children who, at the time of investigation, were aged 6–9. A 2 ml blood sample was taken from the veins of those children whose parents gave their consent. Blood lead levels ranged from 2.9 µg/dL to 34.0 µg/dL.

Neuropsychological performance was assessed by using a selection of tests for ability and attainment drawn from the British Ability Scales. Potentially confounding variables were chosen by reference to '...long-term child development studies, reports in psychology journals and findings in other lead studies.'[67] The children were not from socially deprived backgrounds, and the class composition of the sample closely matched that of the background population from which they were drawn.

Fulton et al have summarized their results by saying

A multiple regression analysis was made with test score as the dependent variable, log blood-lead as the independent variable, the potential confounding variables as covariates, and school as a factor. There were statistically significant negative relations between log blood-lead and all three test scores, after adjustment for all the covariates...For all three scores the relations between lead and test scores were stronger before adjustment for covariates.[68]

The results of this study, in respect of PbB measurements, are illustrated in Figure 3.1. The horizontal axis indicates blood lead levels using both a normal linear scale and a logarithmic scale. The vertical axis shows ability scores above and below the average. The dotted horizontal line across the centre of the illustration represents the null hypothesis that ability scores are unrelated to body lead loads. The dark dots represent the average ability scores for various blood lead levels, and the small vertical lines set over the dark dots represent the 95 per cent confidence intervals for those estimates. The dark line which slopes from upper left to lower right represents the best line of fit to the dark dots. It clearly indicates that, on average, the higher the PbB level the lower the ability score.

When the UK's Medical Research Council reviewed this study their committee commented that

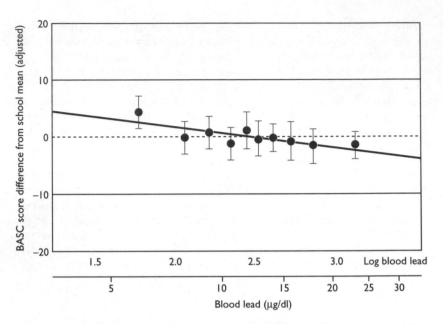

Figure 3.1: *Ability score (means and 95% confidence intervals) for 10 groups of children ordered by blood lead*

Note: The ability score measures the difference from school mean and is adjusted for confounding variables

Source: M Fulton et al (1987) 'Influence of blood lead on the ability and attainment of children in Edinburgh' Lancet 30 May, p1223, Fig 7

> *The design of this study is generally sound, and the sample size provides adequate power. However there remain a number of reservations which must be taken into consideration when interpreting the study results. The sample was drawn from a socially advantaged population...The population studied also differs from those in other UK studies in that water lead was a significant source of body lead.*[69]

That comment is slightly curious and indicates a level of scepticism inflicted on positive findings which was not matched when the committee commented on studies that failed to indicate evidence of an effect. Since the main source of lead for this group was drinking water it would be reasonable to assume that the rate of intake of lead by the children would have been relatively constant, and therefore that the levels of lead in their blood provided a more, rather than less, reliable indicator of their body lead loads than might otherwise have been the case. So while this may have been a distinctive feature, it is one which adds to, rather than detracts from, the reliability of the result.

The MRC committee also remarked that

> *...the potential confounders used [by Fulton et al] were based on prior hypotheses and not on their association with lead or outcomes within this population. While it may be argued that this avoids bias, it could also be argued that they were not necessarily the appropriate confounders since they were based on information from other less socially advantaged populations, and on populations where the sources of lead were dissimilar. It is notable that rather few of the covariates were found to be significantly associated with blood lead measures.*[70]

If the researchers had adopted the opposite tactic when selecting potential confounders they might well have been liable to far more serious criticism that they lacked any adequate prior appreciation of factors which might have confounded their results, and that they had simply 'gone fishing' in the data for related cofactors. An approach along those lines also would have risked over-controlling for possible covariates.

The approach adopted by Fulton and her colleagues was based upon a careful scrutiny of a wide range of other studies, not all of which were conducted with relatively impoverished communities, and thereby avoids the kinds of bias which might otherwise have arisen. The Edinburgh study should therefore be counted as one of the most reliable and robust of all the general population cross-sectional blood lead studies. That was certainly the case at the time of its publication.

The New Zealand study

One of the potentially most powerful of all the general population cross-sectional studies took place in New Zealand in the mid-1980s, and it was based on measurements of lead in teeth.[71] Its power derived in part from the very large sample size, but also from the care taken in other key respects. Deciduous teeth were initially collected from a total of 1035 children, which represented 95 per cent of all the children from the target population; to attain such a high participation rate was a remarkable achievement. Given that some of the teeth were damaged, it was possible to estimate dentine lead levels for only 996 teeth. The children were tested psychometrically when they were between eight and nine years old, using the Wechsler Test of Child Intelligence (known as WISC). The average levels of lead found in their teeth was, on average, about half the corresponding levels that had been reported by Needleman et al in their 1979 paper, namely 6 μg/g by comparison with the figure of 14 μg/g found in the US study.[72]

At the end of the study a complete data set, comprising dentine lead values, IQ and word recognition tests and a rich set of potential covariates, had been obtained on 724 children aged eight. Groups of a very similar size were available at age nine, and with estimates of closely related key variables. Following their initial statistical analysis, the authors concluded that they had been able to show the presence of small but consistent statistically significant correlations between dentine lead levels and test scores, across the entire range of tooth lead measurements which they had found, namely from over 20 μg/g down to below 1 μg/g.[73] The results of this study were analysed in an unusually complex and sophisticated fashion, and the investigators' overall conclusion was that '...nearly all measures relating to the child's school performance showed small but statistically significant associations with lead values even after control for sources of confounding.'[74] The changes which the investigators found in what they refer to as 'measures' of intelligence ceased, however, to be statistically significant once adjustments were made for the selected covariates.

Cross-sectional studies: a summary

The position, therefore, towards the end of the 1980s was that a reasonably large number of studies on the relationship between childhood body lead loads and neuropsychological functioning had been conducted, but their results were equivocal and inconclusive. That was not, however, the end of the matter. Two further approaches have subsequently been developed. One was to apply the techniques of what statisticians have come to call 'meta-analysis', while the other was to conduct and report the results of long-term prospective studies. These two approaches are discussed in turn below.

Meta-analyses of the results of cross-sectional studies

One of the simplest ways of trying to come to terms with the diversity of results from individual studies on the possible link between performance in IQ tests and body lead loads would be simply to list all the studies and then to indicate those which did, and those which did not, provide statistically significant evidence of an effect. It has become customary to count a result as statistically significant if the evidence suggests that there is less than one chance in 20 of the result having occurred at random, and that condition is represented by statisticians by saying that the value of *P* is below 0.05.

As Needleman and Gatsonis have argued, however, such an approach would give undue emphasis to the individual study's P value and would attach equal weight to all studies without regard to their specific merits or flaws.[75] The other factors which should be taken into account include, for example, the details of the studies' methodology and protocols, the size of the sample investigated and the ability of the study reliably to detect a small effect. It is a basic fact of statistics that the information-content that can be derived from a study rises with the square root of the sample size, other things being equal.

There is a systematic way in which many of those considerations can be taken into account in a rigorous mathematical fashion. Called a 'meta-analysis', the technique has been applied to studies of lead and health at least three times, firstly by Needleman and Gatsonis,[76] secondly by Winneke et al[77] and thirdly by the World Health Organisation.[78] Those three analyses will be discussed in turn. In a meta-analysis the summary results of a multiplicity of studies are pooled by treating them as if each of them were providing a set of data points in a larger meta-study. The main advantage of developing a meta-analysis is that it enables the effective sample size of the meta-study to be increased to the sum of the sizes of the individual samples in the separate studies.

Before a meta-analysis can reliably be used as a investigative tool, certain conditions need to be met, but once those conditions are satisfied, a meta-analysis can provide a very powerful statistical tool.[79] The main requirement for the application of the technique is that the relevant studies are similar in several key respects. In particular, the groups between which comparisons are being drawn should be sufficiently similar, essentially the same set of variables should have been investigated across the set of studies, and the studies should be similarly capable of detecting the effects for which the investigators had sought.[80]

The meta-analysis by Needleman and Gatsonis

Needleman and Gatsonis thoroughly searched the published literature and located a total of 27 studies of the effects of lead-exposure on children's performance in IQ tests which had been reported between 1972 and 1990, which was when Needleman and Gatsonis published their analysis. Of those 27 studies, the authors concluded that it would be possible to conduct a meta-analysis on just 12 of them. Studies were excluded from the meta-analysis for several reasons: for example, some did not adequately control for covariates reflecting socioeconomic factors; some included subjects who had already been diagnosed as having clinical lead poisoning; in some cases there were insufficient data to permit proper quantification or to permit the calculation of what statisticians would refer to as the coefficient of lead in a multiple regression model; and one study was excluded on the grounds that it had over-controlled for lead exposure.[81] The studies included in the meta-analysis by Needleman and Gatsonis are listed in Table 3.2.

Needleman and Gatsonis then divided those 12 studies into two groups: those based on tooth lead levels and those which had measured blood lead levels, and these two subgroups were checked for what is known technically as 'statistical homogeneity', that is to say that they were similar in several quite precise respects.

When they had completed their analysis, Needleman and Gatsonis reported a combined P value for the group of studies which used blood lead data of less than 0.0001, which means that there is less than 1 chance in 10,000 that the effect had occurred randomly. For the group of studies which used levels of lead in teeth, one technique produced a *P* value of less than 0.005 while the other was below 0.004.[82] These figures are very revealing because they show that while a simple tabulation would suggest that the evidence taken collectively was equivocal, the meta-analysis indicates that if each small study had been a part of one larger study then a very clear cut result would have emerged.

Needleman and Gatsonis were then able to reinforce that conclusion by following a procedure in which the sensitivity of their analysis was scrutinized by removing each study from the calculations one by one, and then recalculating the combined estimates of statistical significance for those which remained.[83] The conclusion which those recalculations point to is that no single study seemed to be responsible for the significance of the overall result.[84] Needleman and Gatsonis were, furthermore, able to show that the criteria which they had used to exclude 12 of the initial set of 24 studies had not biased their meta-analysis.[85]

Needleman and Gatsonis also considered the possibility that some studies had not been reported, or that journals had declined to publish studies. This possibility was worth investigating because the editors of academic journals are sometimes less keen to publish negative results than positive ones. Needleman and Gatsonis therefore calculated '...the number of unpublished nonsignificant studies that would be necessary to

Table 3.2 *Studies included in the meta-analysis by Needleman and Gatsonis*

Study	Year	Exposure Measure	Subject's age in years	Country
Yule et al	1981	Blood	6–12	United Kingdom
Lansdown et al	1986	Blood	Preschool	United Kingdom
Winneke et al	1983	Tooth	7–12	Germany
Needleman et al	1985	Tooth	7–8	United States
Ernhart et al	1985	Blood	Preschool	United States
Schroeder et al	1985	Blood	1–6	United States
Hawk et al	1986	Blood	3–7	United States
Fergusson et al	1987	Tooth	8–9	New Zealand
Fulton et al	1987	Blood	6–9	United Kingdom
Hatzakis et al	1987	Blood	7–12	Greece
Pocock et al	1987	Tooth	6	United Kingdom
Hansen et al	1987	Tooth	7–8	Denmark

Source: H Needleman and C Gatsonis (1990) 'Low-Level Lead Exposure and the IQ of Children, A Meta-analysis of Modern Studies' *Journal of the American Medical Association* Vol 263, No 5, 2 February, p675

bring the overall *P* value to greater than 0.05.'[86] Using one approach to that problem they calculated that 26 null result studies would have been necessary to dilute the finding for the tooth lead group to that degree of statistical insignificance, and that 67 studies would have been required to dilute the blood lead result. Using an even more stringent procedure, they concluded that '...it would require 16 and 35 studies to dilute the finding for the tooth lead group and the blood lead group, respectively.'[87] Needleman and Gatsonis then rather dryly observe that 'Given the expense of conducting human studies of lead exposure and the amount of attention directed to this question, it is unlikely that this number of negative studies have escaped notice.'[88] That remark must be interpreted as an ironic understatement.

As Yule and Rutter have observed, '...there are powerful lobbies on both sides only too willing to promulgate research of any quality that supports their case, and it is therefore highly improbable that any significant research remains unknown.'[89] We can be entirely confident, for economic rather than scientific reasons, that negative studies, that is studies which failed to detect an adverse neuropsychological effect of low levels of lead, would have reached the public domain.

Needleman and Gatsonis therefore conclude that 'The overall evidence...establishes a strong link between low-dose lead exposure and intellectual deficit in children.'[90] They were able to make that claim because the PbB levels over which that result held ranged from below 10 μg/dL to above 30 μg/dL. Their analysis therefore provided stronger evidence than that from any of the individual studies included in the meta-analysis.

Winneke et al's meta-analysis of the EC/WHO study
In an attempt to contribute to a resolution of the long-running debate about the effects of lead on children's health, the Commission of the European Community sponsored a multi-centre study in Europe in collaboration with the World Health Organisation in the mid-1980s.[91] Eight separate studies were conducted: at Sophia in Bulgaria, Aarhus in Denmark, Athens in Greece, Budapest in Hungary, Modena in Italy, Bucharest in Romania, Düsseldorf in Germany, and Zagreb in the Croat region of what was then Yugoslavia. To facilitate a collective analysis of the data from all eight centres, a common protocol was developed. Four of the eight studies (namely those in Sophia, Bucharest, Düsseldorf and Zagreb) were smelter studies, while the remaining four were studies of the general population of those urban areas.

In the event, however, '...only five groups took part in the quality control programme for tooth-lead analysis using ground tooth powder with known concentrations provided by the [chief investigator] in Düsseldorf ...The initial plan called for using both blood-lead and tooth-lead concentrations as markers of environmental lead exposure.'[92] However, in the course of the study, it became clear that it was difficult to collect a sufficient number of teeth for analysis, and a decision was therefore taken to use only blood-lead levels as markers of exposure.

The investigators explained that their choice of outcome measures was not guided by theoretical considerations relating to the effects of low-level lead exposure on neurobehavioural function, instead the selection was made by reference to those previous studies which seemed to reveal a definite effect. Consequently, IQ estimates were made using the WISC scales (but not the more reliable and up-to-date revised version which is usually referred to as WISC-R), supplemented by tests for visual-motor integration, reaction performance and general behaviour ratings. The data were analysed by a team based in Düsseldorf using multiple regression analysis.

With the exception of the Zagreb study, all the studies revealed some significant adverse relationships between blood lead loads and some of the behavioural outcomes. In relation to the test of visual-motor integration, the results indicated a consistent adverse trend, but they reached statistical significance only in the case of the population close to the Sophia smelter plant. In relation to the results of the WISC tests, there was a marked variation between the eight different groups. In only five of the eight studies was there a negative correlation between PbB levels and WISC scores, and only one of those five reached statistical significance. Winneke and his colleagues account for this by saying that 'These systematic differences are mainly due to inappropriate age norms of the old WISC as opposed to the revised form (WISC-R), as well as the lack of culturally adapted normative data in some of the participating countries.'[93]

This recognition on the part of the investigators is particularly important. For reasons which were discussed in Chapter 2, enormous problems of interpretation can arise if you try to use an IQ test developed for use with one particular social group on a society which is quite different. The fact that the WISC test system was not applied in a variety of culturally adapted forms in the different European societies implies that trans-national comparisons have a strictly limited utility.

Winneke et al concluded that 'There is a weak negative association for environmental lead exposure and psychometric intelligence which is of only borderline significance... Stronger and more consistent associations with lead exposure exist for disruption of visual-motor integration...and of serial choice performance.[but]...No significant effects exist for less standardized measures of neurobehavioral functions.'[94] Indeed they remarked that 'Neurobehavioral effects of environmental lead exposure in children represent weak signals embedded in a noisy background.'[95]

The meta-analysis conducted by the International Programme on Chemical Safety of the World Health Organisation
In 1995 the World Health Organisation (WHO) published its long-awaited report on lead.[96] When preparing their report, the WHO's working party of the International Programme on Chemical Safety (IPCS) conducted a meta-analysis of ten of the published cross-sectional studies on the relationship between PbB levels and scores on IQ tests. They were selected by reference to the care and thoroughness with which they had been conducted and reported, and they included the eight studies covered by the analysis of Winneke and his colleagues, plus three others, namely those from Lavrion in Greece, Edinburgh and London. The WHO-IPCS team selected one primary indicator with which to integrate their analysis; they calculated the

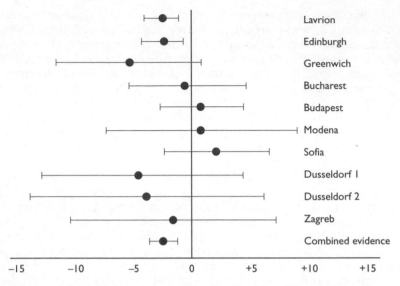

Estimated mean change in IQ for an increase in blood lead level from 0.48 to 0.96 μmol/litre (10 to 20 μg/dl) in cross-sectional studies

Figure 3.2: *The meta-analysis of ten cross-sectional studies by the WHO's ICPS*

likely adverse impact on IQ scores which would result from an increase in blood lead levels from 10 μg/dL to 20 μg/dL. They provided a simple illustration of the results from each of these ten studies, and a pooled analysis in a graphic display, which is reproduced above as Figure 3.2.

The line which represents the combined evidence from all ten of the studies provides a strong indication that an increase in PbB levels from 10 to 20 μg/dL is correlated with a deficit of approximately 2 IQ points, and there is substantially less than 1 chance in 20 that the result had occurred randomly.

Meta-analysis: a summary

The contrast between these three meta-analyses might seem puzzling because only two of them yield a clear cut result, but in practice they all point in the same direction; it is just that the first and the third do so more strongly than did that from Winneke and his colleagues. The position which all of this work produced, therefore, was strengthening evidence of a strong correlation between elevated body lead loads and poor performance in neuropsychological tests, and at progressively lower levels of exposure, down to and even below 10 µg/dL. The challenge which always could, and frequently was, issued remained, however: is the link a causal one, or is the correlation purely coincidental? And if there is a causal link, which is the cause of which? Is lead damaging the children's mental performance, or are the children absorbing more lead because they are not particularly smart? It is this issue which will now be addressed.

Is the link a causal one?

At the conclusion of their discussion, Needleman and Gatsonis addressed the critical question of whether the link, which they believe they have established between elevated levels of lead and IQ decrements, is causal.[97] They acknowledge that cross-sectional studies cannot by themselves resolve the issue of causation, but they contend that the hypothesis of a causal link is strongly supported by several considerations. These include, firstly, the fact that there is biochemical evidence, primarily from animal models, indicating well-defined mechanisms which explain the effects which have been detected. Secondly they argued that there is no evidence that the adverse results can be explained by reference to any single confounder or set of confounders. They pointed, thirdly, to the fact that the adverse consequences of lead exposure have been demonstrated consistently in numerous studies under similar circumstances.[98] Those considerations make a causal link more likely, but they do not demonstrate the link. In the 1980s scientists increasingly came to acknowledge that the most effective way of establishing causality would be by conducting what are called 'prospective longitudinal' studies, namely studies which follow children as they develop, to see whether correlations can be found between their accumulated lead loads and their subsequent performance and development. The question therefore became, could evidence be found that body lead loads are an antecedent cause of the children's later performance?

General population prospective and longitudinal studies

A conference was held in Cincinnati in September 1981 which reviewed the design and conduct of studies of the link between body lead loads and the mental development of children.[99] A consensus was reached that problems concerning the measurement of covariates and confounders, as well as better documentation of the lead exposure variable and exposure histories, could best be addressed by undertaking several prospective studies.[100] Such studies would start by identifying a sample population group and would then follow their development. They could start from the prenatal period and continue through birth and subsequent development, monitoring blood lead levels at each stage, and exploring longitudinally the possible relationships of PbB levels with the development of those children.

A second conference was held in 1984, again in Cincinnati, and several groups of scientists then reported that such studies were being planned. Of those, five have subsequently been conducted and reported.[101] All of this new generation of studies were designed in a similar way. They have all provided data on sequential blood lead levels at regular intervals at least until the children reached seven years of age, as well as data on performance in neuropsychological tests. Each of these studies employed multivariate analysis to try to take account of possible confounding factors upon their measures of neuropsychological outcome.

Although it was generally understood that longitudinal studies might be able to provide distinct advantages over cross-sectional studies, it would be wrong to suppose that they pose no problems whatever. One of the main problems is that of retaining the members of the initial sample for the entire duration of the study. This is especially important as the duration of the study lengthens.[102]

In the course of longitudinal studies, neuropsychological performance is tested at regular intervals, but as children develop they go through several quite distinct stages. The neuropsychological performance tests should not be thought of as if they were measuring a single property or constant characteristic as it moves though different stages. It would be more realistic to think that different characteristics are being assessed at successive stages, even though there is a presumption that what is being assessed possesses some underlying continuity.[103] It almost goes without saying that a six-month-old baby can do quite different things from the actions characteristic of a two-year-old child. Researchers might be seeking some standardized and normalized scale with which to estimate how children are developing by comparison with the average for their age, but they are not tracking the development of one single property.

A fascinating contribution was made to lead toxicology in 1984 which reinforces this point. Shaheen conducted a very intense scrutiny of the development of a relatively small group of children; her final results were based on a sample of only 18 children between the ages of four and six and 18 matched controls; but her insight is intriguing.[104] Shaheen provides evidence indicating that the adverse neuropsychological effects of lead in young children vary depending on the stage(s) at which exposure occurs. This is important because it means that it would then be unrealistic, and would have been unrealistic for all these years, to have looked for a uniform pattern of symptoms unrelated to age or the duration of exposure. Shaheen argued that the adverse effects of lead poisoning on a baby in the first 18 months of its life, when it was developing sensory-motor skills, would differ from those which would occur during the next 18 months, when children are learning to use language.[105] The results which she produced were not conclusive, but they draw attention to an important point.

Most studies of lead neurotoxicity have tended to assume that they were exploring the relationship between two similarly determinate characteristics, but that confidence now seems to have been misplaced. Whenever levels of lead in the blood are estimated, pretty much the same property is being studied, but neuropsychological performance at age one is different from that at age four, and both are different from that at eight. If, therefore, the harm that lead is exerting changes with age and circumstances, tests to estimate the magnitude of that harm are more likely to have underestimated rather than overestimated it, unless exceptional care was taken to use only those tests which had been carefully calibrated over a wide age range. So the evidence that we have from cross-sectional studies on children of various ages deserves to be assigned greater weight than we might otherwise have supposed because that calibration has inevitably been imperfect.

Five prospective and longitudinal studies

In all of the five major prospective studies, body lead loads have been estimated by measuring blood lead levels, although in Cleveland, Ohio, teeth were also collected and analysed. The five studies have been located in Boston, Massachusetts; Cincinnati and Cleveland, both in the State of Ohio; and at two Australian sites, namely Port Pirie and Sydney. In each of those studies, samples of blood were taken from mothers, prior to the birth of their children (known as 'prenatal' samples), from newborn babies (postnatal samples) and from children. Umbilical cord blood samples were often also taken.

One unanticipated benefit of prospective studies has, however, recently emerged – a new index of lead exposure has been developed, and at least partially validated. It was first referred to by scientists conducting a prospective study in the Australian town of Port Pirie.[106] It has come to be known as a child's lifetime average blood lead level, and it can be readily defined and calculated. The measure is, as its name suggests, a statistic calculated by averaging across all of the separate PbB measurements that were taken at regular intervals; the units remain as µg/dL. This indicator can tell us more about a child's overall body lead load than an instantaneous measurement because

> ...*peak blood lead level, (usually observed during the second year of life) is considerably greater than the average lifetime blood lead. For example, a 5 year old child with a lifetime average blood lead of 20 µg/dL may have experienced blood lead levels above 40 µg/dL during the second year of life and may have spent 3 years of life, from 12 months to 48 months, above 20 µg/dL...A lifetime average blood lead of 15 µg/dL should not be interpreted as a single point in life, rather it implies sustained exposures, substantially above 15 µg/dL*[107] *(emphasis added).*

This is particularly important because it implies that, in so far as lead is neurotoxic, one could expect, on average, a greater adverse effect on neuropsychological performance from a child with a lifetime average PbB figure of 10 µg/dL than from a child with a PbB value at that figure on a single occasion.

Another interesting result, emerging from the recent review of the results of the five major longitudinal studies, has been to confirm the suspicion that the relative value of single blood lead measurements is a function of the routes of exposure. As the participants at a major international meeting concluded,

> *The shape of the longitudinal blood lead profile is dependent on the primary source of exposure. For example, exposures to lead in dust result in elevations in blood lead beginning about 6 months of age and peaking between 15 and 30 months of age. If the source of lead is drinking water, high blood lead levels may be present throughout both the foetal and early neonatal periods.*[108]

This result implies, for example, that the results of the Edinburgh study discussed above should perhaps be treated as rather more reliable than those of some other studies which examined children whose main source of lead was, for example, from food or paint and dust.

The five main prospective studies are discussed below, taken alphabetically, namely 1. Boston, 2. Cincinnati, 3. Cleveland, 4. Port Pirie and 5. Sydney.

The Boston prospective study
Several venerable veterans of lead toxicology, including Herbert Needleman and David Bellinger, have been responsible for the organization, conduct and reporting of the Boston-based prospective study.[109] Their initial sample consisted of 11,837 babies, and for each of them an umbilical cord blood sample was taken and analysed.[110] By reference to those measurements, they were divided into low, medium and high-lead groups. The low-lead group had cord PbB levels below 3 µg/dL, with an average of 1.8 µg/dL; the medium-level group had cord PbB levels between 6 and 7 µg/dL with a mean of 6.5 µg/dL; while the high-lead group had cord PbB values above 10 µg/dL, with a mean figure of 14.6 µg/dL.[111]

The study then focused on a selected cohort of 249 babies drawn from mainly middle- and upper-middle class families. Infants were excluded from the study if, for example, they suffered with an independent risk factor for difficulties in mental development, such as Down's syndrome, if their first language was not English, or if maternal consent was withheld.[112] By the time the infants had reached two years of age, the size of the cohort had declined to 182 children. The relatively affluent background of these children was consistent with their relatively high performance on cognitive tests. The mean adjusted Mental Development Index (MDI) scores at 24 months, for example, were 119 for the low-lead group, 118 for the middle group and 111 points for the high-lead group, as compared with an overall population average of 100.[113]

Samples of blood were obtained by capillary tube from finger pricks at six-monthly intervals, rather than by venipuncture. A distinctive feature of the sample is that, on average, infantile PbB levels were both low and relatively stable between 6 and 24 months. This was probably a consequence of the relative affluence of the neighbourhoods in which they were living, but it is significant because it facilitated the separation of potential effects of prenatal from postnatal exposures. When postnatal PbB levels are found to be higher, as they were for example, in Port Pirie and Sydney, then it becomes far harder to disentangle effects from the two distinct routes of exposure.

The investigators found a clear inverse relationship between MDI scores and prenatal PbB levels at 12, 18 and 24 months. They reported that, averaged over the first two years of life, for every 10 µg/dL increase in average postnatal PbB levels, MDI

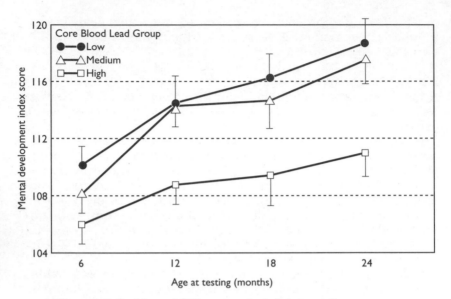

Figure 3.3: *Mean MDI scores in infants at four ages as a function of their umbilical cord blood levels*

Note: Scores are least-squares means obtained by regressing MDI scores on the cord blood lead group and 12 variables considered to be potential confounders. Error bars represent 1 SD, but for clarity bars extend only in one direction.

Source: D Bellinger et al (1987) 'Longitudinal analyses of prenatal and postnatal lead exposure and early cognitive development' *New England Journal of Medicine*, 23 April, Vol 316, No 17, p1041

scores declined by something between 4.6 and 6.1 points. The mean adjusted MDI scores for the three groups over the first two years, as a function of age and cord blood categorization, are illustrated in Figure 3.3.

By the time the children had reached 57 months, ie four years and nine months, blood samples were collected by venipuncture, and a difference remained in average GCI scores (or General Cognitive Index from the McCarthy Scales of Children's Abilities) between the high and low cord PbB groups, but the differences by then were smaller, and they were not statistically significant. This could be interpreted as indicating that the cohort had recovered, at least partially, from such adverse effects of lead as might previously have occurred.[114] By that stage, the mean PbB level for the remaining cohort had declined to 6.4 µg/dL, while the highest level found was 23.3 µg/dL.[115]

The lead industry pressure group ILZRO interprets these results as implying that what they term 'environmental enrichment' '...can facilitate the rate of recovery between the ages of 24 and 57 months, after which the effect disappears.'[116] It may be that the relative richness of the diets, experience and education of these children was sufficient to swamp any adverse effects which prenatal lead exposure might have had, but that suggestion should be treated as a hypothesis, not as a robust conclusion. In this context, as in many others, the absence of evidence should not be interpreted as evidence of absence.

By 1992, however, the Boston team was in a position to report on its follow-up investigation of 148 of the original cohort of children, who had reached ten years of age.[117] The investigators examined two important neuropsychological indicators, namely the performance on the Wechsler Revised Intelligence Scale for Children (WISC-R) and the Kaufman Test of Educational Achievement (K-TEA).[118] Bellinger and his colleagues reported that 'Higher levels of blood lead at age 24 months, but not at other ages, were significantly associated with lower global scores on both the WISC-R and the K-TEA after adjustment for potential confounders.'[119] Moreover, 'Over the range of approximately 0 to 25 µg/dL, a ...10 µg/dL increase in blood lead at 24 months was associated with a 5.8-point decline in WISC-R Full-Scale IQ...and an 8.9-point decline in K-TEA Battery Composite Score...'[120] This is, perhaps, a rather striking result because it overwhelms the previous suggestion that effects which were be seen at any early stage were subsequently eliminated before the children reached the age of six. That conclusion is reinforced by the fact that PbB levels typically peak around the age of two years, and therefore exposure at that age is more likely to generate an effect than exposure at other times, at least when dust rather than drinking water provides the dominant source of lead exposure.

Indeed Bellinger et al have observed that

> It is unclear whether this reflects a special vulnerability of the nervous system during this period or simply the fact that blood lead level tends to peak in the second year. Our finding that pb24 [ie PbB at 24 months] was more predictive of performance than was maximum blood lead level supports the hypothesis of an age-specific vulnerability.[121]

The researchers acknowledge that, since they had not been able to follow-up 100 per cent of their sample, there is a risk that the sample which they were able to contact and test was unrepresentative of the larger cohort. They were able, however, to provide evidence that, by reference to earlier data, the subsequent participants and non-participants were not statistically distinct, at the earlier stage of their investigation.[122]

The Cincinnati prospective study
Mothers and children were recruited to the Cincinnati study between 1979 and 1984, mostly from poor inner city house-holds.[123] Prenatal maternal blood samples were obtained from 261 women, while neonatal and subsequent blood samples were obtained from 297 babies, 256 of whom (ie just over 86 per cent) were black.[124] Most of the blood samples were collected by venipuncture, although as the researchers explain, '...finger stick and heel stick methods were used when the physical or behavioral characteristics of the infant demanded it.'[125] During the first phase of this study, the mental development of the sample of infants was assessed, in the first place, at 3, 6, 12 and 24 months, using the Bayle Scales including the MDI.[126]

The mean MDI score at three months was close to the average of the entire population, at 100.38. By 12 months the figure had risen to 111.89, but by 24 months it had declined sharply to 88.08.[127] The researchers commented in 1990 that 'The rather dramatic drop at 2 years is not unusual for lower socioeconomic status infants and probably reflects the relatively greater number of items at this stage that require a verbal or nonverbal response to representational stimuli, rather than sensorimotor manipulation.'[128]

The investigators found that for the mothers of male babies (but not the females), prenatal PbB levels were significantly related to 6-month MDI after adjustment for confounders. The infants' neonatal blood lead levels were also inversely related to MDI scores at six months after similar adjustment, and that relationship was particularly marked amongst the boys.[129] When it came to the 24-month tests, however, the researchers reported that 'No statistically significant relationships between prenatal and postnatal blood lead level variables and Bayle MDI were found.'[130] This was, however, contrary to the researchers' expectations.

One possible explanation for the failure of any effect to emerge at that stage was that, although many of the children came from poor inner city black households, they were benefit-ing from some relatively effective mother and child feeding and nutrition programmes, and may therefore have been better able to withstand the impact of elevated lead levels than might other-wise have been the case.

A cohort of 258 children was subsequently tested once they reached the age of four years, and the results reported in 1991.[131] The researchers then reported that

*The Kaufman Assessment Battery for children (K-ABC)
was administered...Higher neonatal PbB levels were
associated with poorer performance on all K-ABC
subscales. However, this inverse association was limited
to children from the poorest families. Maternal PbB levels
were unrelated to 4-year cognitive status. Few statistically significant associations between postnatal PbB
levels and K-ABC scales could be found. However, the
results did suggest a weak inverse relationship between
postnatal PbB levels and performance on a K-ABC
subscale which assesses visual-spatial and visual-motor
integration skills.*[132]

By 1992, Dietrich et al were able to report on the results of tests
conducted with 259 children who had reached the age of five.[133]
At that stage, they found that 'Higher postnatal PbB levels were
associated with poorer performance on all cognitive development subscales of the Kaufman Assessment Battery for Children
...However, following adjustment for measures of the home
environment and maternal intelligence, few statistical or near
statistically significant associations remained.'[134]

The results of testing the Cincinnati cohort at the age of six
years were published in 1993, and were analysed on that
occasion by reference to their lifetime average blood lead concentration, rather than by any single PbB estimate.[135] As Mahaffey
explained, 'The longitudinal evaluation of severely disadvantaged
children in Cincinnati identified...an eight-point decrement (from
91 to 83) in the performance IQ (based on WISC-R scores after
adjustment for covariables) as the lifetime average blood lead
concentration increased from 10 to 35 µg per decilitre.'[136]
Similarly Dietrich and his colleagues stated that 'The difference
in IQ between the lowest and highest exposure groups was
approximately 9 points prior to statistical adjustment for covariates and 7 points following adjustment. It is also evident that the
group with lifetime exposures in excess of 20 µg/dL were the
most compromised.'[137] These results are represented graphically
in an illustration reproduced here as Figure 3.4.

The Cleveland prospective study
The Cleveland study was initiated with a sample of 215 infants
for whom maternal or umbilical cord blood data were available,
but by the time they reached the age of five the sample had
declined to 156.[138] The team conducting this study has included
Claire Ernhart, a long-standing protagonist in debates on lead
and health; but she is untypical in that she has from time to
time worked relatively closely with the lead industry, and has

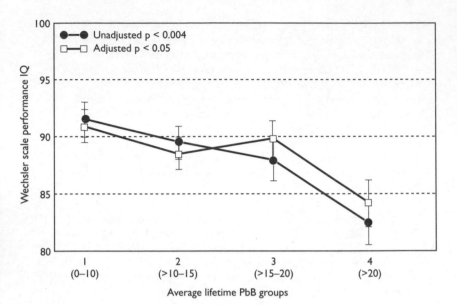

Figure 3.4: *Performance IQ plotted against average lifetime blood lead levels for 253 children aged 6¹/₂ in Cincinnati*

Source: K N Dietrich et al (1993) 'The Developmental Consequences of Low to Moderate Prenatal and Postnatal Lead-Exposure – Intellectual Attainment in the Cincinnati Lead Study Cohort Following School Entry' *Neurotoxicology and Teratology,* Vol 15, No 1, p42

been a regular critic of the work of Herbert Needleman and his colleagues in Boston.

The sample came from families, most of whom were relatively disadvantaged, one-third of whom were black.[139] As a reviewer has explained,

> *Many of the mothers* [whose children comprised the sample] *were alcoholics, having been chosen originally for participation in a larger study of the prenatal effects of alcohol. In keeping with their disadvantaged upbringing, they lost IQ points with age. Mean MDI scores at 6 months, 1 year and 2 years were 112, 112 and 102 points. The mean ...[Stanford-Binet IQ]...score at 3 years was 90 points.[140]*

The average maternal prenatal PbB level was relatively low at 6.5 µg/dL, while mean cord PbB level was 5.8 µg/dL.[141] Postnatal blood lead levels rose quite rapidly, as might be expected with poor inner city children. Mean levels rose sharply from 10.1 µg/dL at six months to 16.7 µg/dL at two years. After testing the children when they were three years old, the investigators reported that there was evidence of a relationship between lower scores in mental development tests and high PbB figures at six months, one year, two years and three years. When adjustments were made for the presumed covariates and confounders, however, the only correlation to remain statistically significant was that for maternal, prenatal PbB levels.[142]

In the Cleveland study, not only were blood samples collected and analysed, but tooth samples were also taken, dentine lead levels were estimated, and then compared with IQ levels for children who had reached the age of four years and ten months, for 164 children.[143]

In 1993 Green and Ernhart provided an analysis of the relationship between dental lead levels and children's performance on IQ tests.[144] Their analysis was however presented in an unusual form. They invoked a rather unusual indicator of cumulative body lead loads. They did not provide analyses in terms of the two most commonly used variables, namely the raw tooth lead data or their logarithmic transformations; instead they provided analyses based on the square roots of the levels of lead found in the children's teeth.

It is not hard to appreciate why they might have chosen this option. The distribution of body lead loads does not have the shape of what statisticians call a 'normal' or bell-shaped distribution. It is not evenly distributed about the average, because typically there is a relatively small number of children with high lead levels. The simple average of the group would therefore represent an overestimate of the levels found in most children. If you transform the blood or tooth lead figures by using the logarithms of the raw numbers, you obtain a pattern which looks more like a bell-shaped distribution, what statisticians thinks of as 'normality', but the transformed distribution has what is termed a 'geometric' mean which tends to underestimate the levels of most of the children. As Greene and Ernhart explain, the square root provides an intermediate between the raw numbers and their logarithms.[145]

Greene and Ernhart reported that, prior to adjustment for covariates, the square roots of the dental lead levels were significantly related to full scale IQ results, verbal IQ figures, and performance IQ, and the effect size was substantial.[146] When statistical adjustments were made, using an indicator of the quality of the domestic care-giving environment over the preschool period, the association between lead levels and performance IQ was no longer statistically significant, but the relations to verbal and to full scale IQ remained.[147]

A curious feature of their report is that even though they go to unusual care to supplement their analysis with calculations showing how their results would have varied depending on the precision in the initial measurements of lead levels and test performance, they failed to spell out the results which could have been obtained if they had taken the other two common options, ie the raw tooth lead data and their logarithmic trans-formations.

A further limitation of the Cleveland-based study, which was pointed out by the World Health Organisation in 1995, was that the results have not been published in a comprehensive form, nor in a form which allows them to be directively compared with the other four prospective studies.[148] The World Health Organisation's IPCS represented the Cleveland study as one which yielded little or no evidence of a link between lead exposure and mental performance, but IPCS seems to have focused on analyses of blood lead data, and failed to refer to the evidence concerning tooth lead. While the evidence of a link between lead exposure and neuropsychological performance is less strong in the Cleveland study than in the studies conducted at the other four centres, the link to tooth lead levels at Cleveland seems distinctly robust.

The Port Pirie prospective study
Of all the prospective studies, the results of the Port Pirie study are particularly important because the researchers studied the largest group of children. At the age of four, 523 children were enrolled in the study.[149] The families were drawn both from the industrial city of Port Pirie (200 km northwest of Adelaide) and from the surrounding agricultural towns. The community is located downwind of a large and long-standing lead smelter, and has consequently been exposed to high levels of lead in air, soil and dust.[150]

The Port Pirie Cohort Study began in 1979, conducted by Peter Baghurst and his colleagues.[151] Early results showed that approximately one-third of the children had PbB levels above 25 µg/dL, in other words the levels of contamination were worry-ingly high.[152] The initial sample consisted of 723 infants born from 1979 to 1982. This represented about 90 per cent of all singleton live births in that community in that period, which is a remarkably high figure. For the 595 children tested at 24 months, the mean MDI score was 109. The developmental status of 494 children was assessed when they were between seven and eight years of age.[153]

Samples of blood were taken from each mother before deliv-ery, from the umbilical cord at birth, and capillary-blood samples were collected from each child at 6 and 15 months, 2 years, and annually thereafter. All the IQ tests were conducted by the same psychologist, who was unaware of the children's lead status. Given that postnatal blood lead levels were significantly higher than umbilical cord levels, and rose quite sharply, it was difficult for the investigators to distinguish between any effects which may have arisen from postnatal exposure and those consequent on prenatal exposures.

The relationship between umbilical cord data and subsequent performance was examined for 523 babies, and for even larger samples at subsequent stages. By 1988 the researchers reported on the results up to the end of the children's second year, saying that 'The Bayle Mental Development Index (MDI) was negatively correlated with PbB at all ages...The correlation attained statistical significance in both the antenatal and the postnatal periods, but not at the time of birth (maternal and umbilical cord samples).'[154] They concluded, moreover, that 'These results indicate that, with other factors remaining constant, a child's MDI at 24 months will be 1.6 points...lower for every 10 µg/dL rise in PbB at 6 months of age.'[155]

By the time that the children had reached four years of age, the researchers had utilized the McCarthy Scales of Children's Abilities (MSCA), which can be used in children aged from three to seven years, and they reported that 'The blood lead concentration at each age, particularly at two and three years, and the integrated postnatal average concentration were inversely related to development at the age of four...[and]...no threshold dose for an effect of lead was evident.'[156]

In 1992, the researchers reported the results of their examination of the children aged seven. They said 'There was a consistent inverse relation between the blood lead concentrations and scores on all IQ scales....The mean IQ scores differed by 2.7 to 12 percent between the children with values in the highest and lowest quartiles for blood lead concentrations.'[157] Moreover,

> *These results indicate that the inverse associations between blood lead concentration and indexes of development reported earlier for this cohort persists into the primary-school years. The strong correlation...indicates that many of the children who scored poorly initially have not had great improvements in their overall ranking by the age of seven years.[158]*

The researchers illustrated their conclusions with a graph repre-

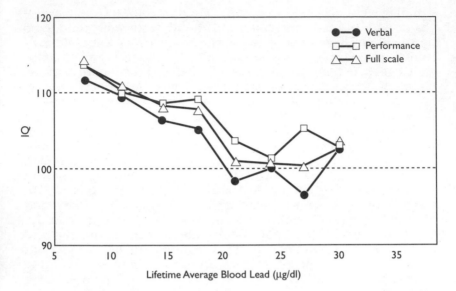

Figure 3.5: *Lifetime average blood lead concentrations and IQ scores at the age of seven amongst children in Port Pirie*

Source: P A Baghurst et al (1992) 'Environmental exposure to lead and children's intelligence at the age of seven years' *New England Journal of Medicine*, 29 October, Vol 327, No 18, p1282, Fig 1

senting three IQ scales as a function of lifetime average blood lead levels, and this graph is reproduced as Figure 3.5.

The graphs in Figure 3.5 illustrate the fact that in this study a consistent pattern was found showing that as average lifetime blood lead levels rose amongst this cohort of children, their scores on the WISC-R scales declined. Their verbal IQ and performance IQ scores, and their overall scores all declined, and that relationship occurred consistently down to PbB levels as low as 5 µg/dL.

Baghurst and his colleagues argued in 1992 that those results were more likely to have underestimated the scale of the problem than to have overestimated it, firstly because the children who dropped out of the study were likely to have had relatively high levels of lead exposure, secondly because they may have over-controlled for social covariates, and thirdly because '... randomly distributed imprecision or misclassification in the measurement of lead exposure (or its confounders) will bias the estimate of the effect towards a null value.'[159]

When the Port Pirie research team published their most recent results in 1996, based on a study of 375 of the children when they were about 12 years of age, the findings reinforced and extended their previous conclusions.[160] For each of the children studied, a lifetime average blood lead level was calculated. The investigators concluded that '...the inverse associations between blood lead and cognitive development at ages 2, 4 and 7 years persisted into later childhood. The estimated deficit in full scale IQ at age 11–13 years was 3.0 points for a shift in lifetime average blood lead concentrations from...10 to 20 µg/dL.'[161] The latest data from Port Pirie also indicate that, with lifetime average blood lead levels ranging from below 5 µg/dL up to more than 30 µg/dL, there was no evidence of a PbB threshold below which no effects seemed to occur.[162]

The investigators also pointed out that 'This study provides evidence that an association between early exposure to environmental lead and cognitive development persists into later childhood, even though blood lead concentrations in these children had declined substantially since their third year of life.'[163] That latter remark is important, because representatives of the lead industry had previously argued that such effects as might have been detected amongst children in the first two years of their lives would have disappeared as they grew older. The latest results from Port Pirie run directly counter to that seemingly over-optimistic claim.

A further piece of the jigsaw was provided on this occasion by an analysis intended to test for the possibility of reverse causation – were elevated PbB levels causing poorer mental performance, or were low IQ scores responsible for higher levels of lead contamination? The investigators remarked that

> *Blood lead measures most strongly related to IQ had all been measured before IQ was assessed and none of the developmental scores at earlier ages were significantly associated with the current blood lead measure at age 11–13 years. These analyses...strengthen the notion that lead exposure influences cognitive development, and not the reverse.*[164]

The Sydney prospective study
The cohort for the Sydney-based study was selected from healthy babies born during 1982 and 1983 in three hospitals, and initially consisted of 318 children.[165] The families were mostly middle class and well educated. Any infant whose mother had drug or alcohol problems was excluded as well as all those whose mothers did not speak English. On average, the quality

of the care-giving environment was good.[166] By the time the children had reached their second birthday, 207 children remained in the study.[167]

Prenatal blood samples were obtained by venipuncture from mothers during their pregnancies, and from umbilical cords at birth.[168] Infant blood samples, up to and including those taken at 24 months, were obtained using both capillary tube and venipuncture techniques, but almost all subsequent samples were taken directly from the children's veins.[169]

When the children were six months old, and at one and two years of age, they were tested using the Bayle MDI scale, but when the children reached the age of three years they were tested using the McCarthy Scales of Children's Abilities, generating both GCI scores and a Motor Scale Score.[170] All neuropsychological examinations of the children were conducted in their own homes within seven days of blood sampling. 'Mean blood lead levels increased from birth to 18 months, [and] then steadily declined to 48 months.'[171] Any adverse effects which the postnatal lead levels might have caused are therefore likely to have swamped such effects as their prenatal exposure might have produced. Consequently, it was difficult for the investigators to detect any effects there may have been from prenatal exposures, but it should have been correspondingly easier for them to detect postnatal effects.[172]

The researchers analysed their data in an orthodox manner, and checked for correlations between performance on IQ tests and lead levels, using both maternal and cord PbB data, as well as postnatal levels. As of 1989, none had been found.[173] In 1995, however, by which time far more data had been gathered and analysed, the World Health Organisation detailed the evidence that significant adverse effects had occurred during much of the first seven years of the children's lives.[174] That report indicates that, at the age of seven, the evidence from the Sydney cohort indicated that a doubling of PbB levels from 10 to 20 µg/dL correlated (after adjustment for confounders) with an average reduction of 1.4 points on the IQ scale, while averaging over the first seven years of the childrens' lives indicated a corresponding mean IQ reduction of 1.6 points.[175]

Summary and interpretation

When viewed as a whole, the balance of evidence emerging from the five prospective studies indicates that, other things being equal, children with higher blood lead levels perform less well in neuropsychological tests than children with lower PbB levels.

There is, moreover, no evidence of a threshold below which those adverse effects cease to occur. As time has passed, the evidence of both a correlation and a causal link has become stronger, and the levels of exposure down to which the effects have been detected has declined steadily too.

In 1995 the World Health Organisation published its long-awaited report on lead.[176] When preparing their report, the working party of the International Programme on Chemical Safety conducted a meta-analysis on the prospective longitudinal studies. They would have liked to have been able to pool the data from all five of them, but the team in Cleveland did not publish their data in a manner enabling their results to be included; consequently only the data from the four others could be pooled. The WHO-IPCS team selected one primary indicator with which to integrate their analysis; they calculated the likely adverse impact on IQ scores which would result from an increase in blood lead levels from 10 µg/dL to 20 µg/dL.[177] They provided, moreover, a simple illustration of the results from each of the four studies, and a pooled analysis in a graphic display, which is reproduced as Figure 3.6. A noticeable feature of this illustration is that it implies that none of the studies by themselves provided a definitive result. This is shown by the fact that for each individual study, the 95 per cent confidence interval crossed the vertical zero line. That means that in each case there was at least a 5 per cent chance that the result was a random aberration. When the data from the four separate studies are pooled, however, and meta-analysed, the larger effective sample produces an altogether more robust conclusion. The

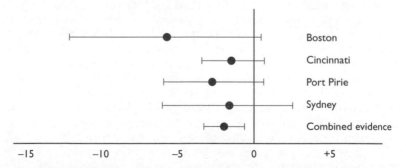

Estimated mean change in IQ for an increase in blood lead level from 0.48 to 0.96 µmol/litre (10 to 20 µg/dl) in prospective studies

Figure 3.6: *A meta-analysis of mean blood lead and full-scale IQ (mean changes and 95% confidence intervals)*

line which represents the combined result from all four of the studies provides clear cut evidence that an increase in PbB levels from 10 to 20 µg/dL results in a deficit of approximately 2 IQ points, and that there is less than a 5 per cent chance that the conclusion had been reached accidentally. While the evidence may fall short of that required to produce total certainty, the evidence is strong enough to satisfy all but the most perverse and recalcitrant members of the scientific community. The argument therefore is no longer whether or not lead is poisonous at very low levels, but focuses rather on the question of what should be done about it.

Since the IPCS is supposedly a purely scientific organization, their report tried to avoid directly addressing issues of public health policy. The IPCS team endorsed the view that 'below...25 µg/dL the size of the apparent IQ effect (at ages 4 and above) is a deficit of between 0 and 5 points (on a scale of 100...) for each 10 µg/dL increment in PbB level, with a likely apparent effect size of between 1 and 3 points.'[178] The report was, however, coy about the interpretation which they placed upon that result. The committee was convinced that, as PbB levels rise above 10 µg/dL, adverse effects occur. What they never quite managed to articulate is the obvious consequence that it is desirable to ensure that PbB levels in young children should remain below 10 µg/dL, and they failed to endorse any particular PbB target.

No such coyness afflicts some senior American scientists. Rosen, who chaired the CDC panel which in 1991 reduced their level of concern to 10 µg/dL, has said that 'The studies in Boston, Cincinnati, and Port Pirie found neurobehavioral deficits in infants exposed to lead at PbB as low as 7 µg/dL.'[179] That finding implies not merely that public health policies should in the first place adopt the target of reducing childhood lead exposures sufficiently to get PbB levels down to 10 µg/dL, but that they should continue to reduce them even further.

The Royal Commission on Environmental Pollution stated in 1983 that 'We are not aware of any other toxin which is so widely distributed in human and animal populations and which is also so universally present at levels that exceed even one tenth of that at which clinical signs and symptoms occur.'[180] If we are to eliminate not just the adverse effects which have so far been found, but also provide a margin of safety, it is therefore necessary not just to cut children's PbB levels to 10 µg/dL, but then to continue to diminish them significantly further.

The evidence which has been reviewed implies that the CDC's 1991 decision to lower their level of concern for children to a blood lead level of 10 µg/dL was robustly grounded. It follows, therefore, that the failure of the British government to respond to the emergence of the evidence and the articulation of such a powerful consensus, represents culpable irresponsibility.

Coda

This chapter has not been a comprehensive review of lead toxicology. It has been a selective discussion of one important strand in the toxicology of lead. Lead is poisonous to children in many other ways, apart from damaging their mental development, and lead appears to damage an even wider range of biochemical functions in adults. For the purposes of this discussion, however, the damage which lead is doing to the intellectual development of young children is especially important. That is partly because children are most vulnerable, but also because much of the evidence of other unwelcome effects of lead has been found, so far, only at PbB levels significantly higher than 10 µg/dL. Adults working with lead are especially vulnerable to lead poisoning, but occupational health and safety is beyond the scope of this text.

4

Lead in the Human Body and the Environment

In 1983, Britain's Royal Commission on Environmental Pollution remarked that 'Because lead has been dispersed so widely as a result of man's activities, there are strictly speaking no uncontaminated sites from which "natural" concentrations in the environment may be determined.'[1] Lead is indeed a very widespread pollutant, but there is evidence that there are a very few places on earth, such as deep in the Antarctic ice, where it is possible to estimate a natural background level which predates human efforts to mine, smelt and utilize metallic lead.

Pattison and his colleagues have used such evidence to help them estimate that past pollution has increased the average lead burden in industrial society to levels approximately 1000 times higher than those experienced by our prehistoric ancestors.[2]

The use of lead as an ingredient in gasoline (petrol, in Britain) has been almost entirely eliminated in the US. As early as 1980, 98 per cent of American cars only used unleaded fuel. In the UK in September 1996 on the other hand, just under one-third of all the petrol sold still contained lead additives. Despite the continued use of lead in petrol in Britain, that problem is at last being addressed. The main sources of exposure to lead in the US now come from the residues of old lead-containing paints, old leaded water pipes and lead in plumbing fittings. The use of leaded fuels in the UK is now declining rapidly, and as it does the relative significance of old

leaded paint and plumbing increases. This chapter will therefore focus on three main areas: levels of lead in human beings,
at least as measured by blood lead levels, and those in their
water supply and old house paint.

Blood lead levels in the US

Providing estimates of blood lead levels has provided employment for an army of scientists and statisticians in the US, while
in the UK it has involved something more like a small platoon.
Many American authorities, at the federal, state and local level,
have diligently endeavoured to estimate the burden of lead being
carried by their populations, and to publish the results of their
work; the same cannot be said for the UK.

Information on blood lead levels, at least in adults, is available for the US going back at least as far as the mid-1930s,
when sampled populations were found to have average PbB
levels ranging between 27 and 34 µg/dL. By the late 1980s,
average adult PbB levels had declined to between 6 and 9
µg/dL.[3] Not all of those sampled will have been entirely representative of US society, and the precision of measurement will
also have been variable. Nonetheless, it is evident that there has
been a long-run declining trend.

There are two main ways of estimating PbB levels in the
population. The first, obviously, is to make direct measurements, which often has been done, but it does necessitate
collecting blood samples which is not always easy, especially
with young children. Over the last 20 years there have been two
thorough official federal studies in the US of blood lead levels in
the American population, sampling both adults and children.
They have been part of the regular National Health and Nutrition
Examination Survey which is commonly referred to by the
abbreviation NHANES. To date, there have been three so-called
NHANES surveys, but lead levels were only estimated for the
last two. The samples for the second NHANES study (known as
NHANES II) were collected between 1976 and 1980, while
NHANES III is far more recent, having taken place between
spring 1993 and spring 1995. The data from these studies
provide the main source of estimates of PbB levels in the
American population.

Modelling environmental and blood lead

The second way to estimate PbB levels in the population is indirect, and involves drawing inferences from what is known about the levels of lead in the environment. This approach has been developed especially in relation to children, from whom it is especially difficult to get blood samples. In the late 1980s the US EPA invested considerable resources into a project to develop a computerized model with which to estimate children's probable PbB levels, by making calculations based upon estimates of the levels of lead to be found in a particular environment, for example in the air, soil, drinking water, house dust and food. The EPA's model was first developed as a risk assessment tool to estimate childhood lead exposures at hazardous waste sites, but it can now be used to generate such estimates across a very wide variety of locations. It has the added advantage that it can also be used to estimate the potential impact of setting and enforcing various standards, for example on lead concentrations in drinking water, food and air. The model has gone through a lengthy process of design and development, and the software that is now available is easy to use and has been well validated. This work was only possible, of course, because a great deal of data had already been collected, mostly in the US, relating PbB levels to those found in the environment.

The US EPA quantitative computer model is known as the Uptake/Biokinetic Model and this is commonly abbreviated to the UBK model.[4] The model is extremely useful because information on levels of lead contamination in parts of the environment, such as air, food, drinking water, soil and dust is often readily available, or relatively easy to obtain, and it can then be incorporated into the model, allowing one to estimate the likely distribution of blood lead levels in a population of children. The model is currently designed to apply to children up to the age of seven years, but there is scope to apply it to older children, and in due course to adults.

The model permits the user systematically to vary levels of exposure, and the parameters which influence uptake and absorption. In the US, it is possible to compare direct measurements of PbB levels with those derived by the EPA's UBK model, and thereby demonstrate the reliability and precision of the model. In the UK, where rather more information is available on levels of lead in the environment than on children's blood lead levels, the UBK model is especially helpful because it can be used to fill what would otherwise be a substantial gap in our knowledge.

The NHANES studies in the US

There is clear evidence to indicate that average blood lead levels in the US have declined, not only since the 1930s, but also over the last two decades. During the course of the NHANES II study in the late 1970s, it became clear that PbB levels had been declining and that that decline had been related to the reduction in the use of lead-based additives in gasoline.[5] That relationship is illustrated in Figure 4.1, which derives from a 1988 report by the Agency for Toxic Substances and Disease Registry (ATSDR).

When the results of the NHANES III study were issued in 1994, the long-run trend of declining blood lead levels in the US population was confirmed, but the results also showed that not all hazardous exposures had been eliminated.[6]

NHANES II
When the NHANES II was conducted, it was the first ever thorough and comprehensive survey of American blood lead

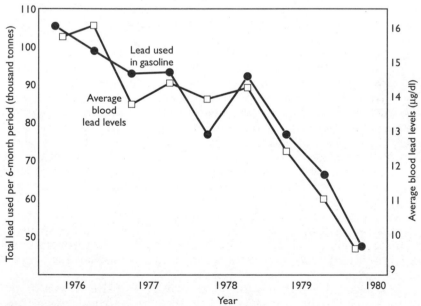

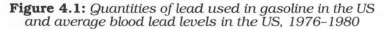

Source: Annest et al (1983)

Figure 4.1: *Quantities of lead used in gasoline in the US and average blood lead levels in the US, 1976–1980*

levels.[7] The blood lead survey component of NHANES II was conducted by the CDC's National Centre for Health Statistics between 1976 and 1980, and blood lead concentrations were determined for 9933 individuals out of a total sample of 27,801 people, aged between six months and 74 years.[8] Blood lead levels were measured in 759 children aged between six months and two years and for 1613 children aged three to five years, and from those measurements mean blood lead levels were calculated for the mid-year of 1979.

The results indicated that those children aged between six months and two years then had a mean blood lead concentration of 16.3 µg/dL, while those aged from three to five years had a mean blood lead level of 15.9 µg/dL.[9] At the time the NHANES II results were published the CDC's 'level of concern' for young children was 30 µg/dL;[10] and at that time 12 per cent of black children aged between six months and five years had PbB levels in excess of that figure, while the same could be said of only 2 per cent of white children.[11]

One important feature of the results of the NHANES II survey was that it showed that blood lead levels in urban and non-urban populations were significantly different. Children aged six months to two years in non-urban environments had mean blood lead levels of 13.9 µg/dL while for those living in large urban areas the comparable figure was 18 µg/dL.[12] The racial contrast is illustrated in Figure 4.2. The vertical axis of this graph represents mean blood lead levels, and the horizontal axis indicates age in years. The two lines tell their own story.

Indeed, the most vulnerable group were black children from low-income families, and the results of NHANES II suggested that almost one-fifth (18.5 per cent) of these children had PbB levels at or above 30 µg/dL.[13] The evidence reviewed in Chapter 3 implies that the mental and intellectual development of these children would have been significantly impaired by their body lead loads, when compared with children from more affluent neighbourhoods.

The results of the NHANES II study showed, moreover, that only 22 per cent of those tested (both adults and children) then had PbB levels below 10 µg/dL. There have been improvements since then, but at that time there was clearly considerable scope for improvement.

Following the publication of the results of the NHANES II study in 1982, there was very little response from either the legislative or the executive branches of the US government, except within the US EPA. Experts within the EPA devoted a huge effort to producing the massive and pivotal 4-volume report entitled *Air Quality Criteria for Lead*. This document

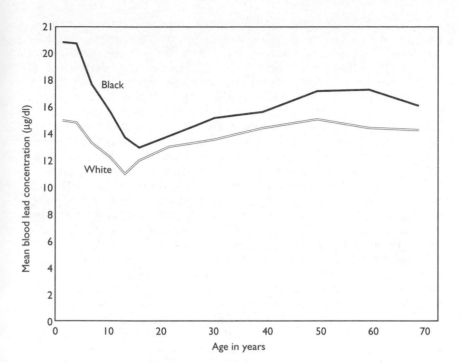

Figure 4.2: *Blood lead levels by race and age in the US according to the Second National Health and Nutrition Examination Survey, 1976–1980*

Source: K R Mahaffey et al (1982) 'National estimates of blood levels: United States, 1976–1980: associated with selected demographic and socioeconomic factors' New England Journal of Medicine Vol 307, Fig 1, p575

explained almost everything that you might then have wanted to know about the characteristics of lead – its sources and emissions, pathways of human exposure, and biochemistry and toxicity, but there was one element missing from the documentation: the EPA provided no estimates of the scale and extent of lead poisoning in the US population. Congress took advantage of this omission from the report which, in all other respects, was comprehensive if not exhaustive. At that stage (1986) Congress was reluctant to impose a complex and costly regulatory system, and delayed taking action, instead instructing the Agency for Toxic Substances and Disease Registry (the ATSDR) to provide a report on the extent, as well as the nature, of lead poisoning in children in the US. That report emerged in 1988, and it provided the next major contribution to the debate about the scale of the problem.[14]

NHANES II had provided evidence from a relatively modest sample of the American population. There were approximately 17 million American children in 1984 between the ages of six months and five years, and NHANES II provided estimates of PbB levels for just 2372 of them, which is approximately 0.014 per cent. The ATSDR transformed the NHANES II data to generate estimates of the proportion of the entire US population with elevated lead levels, as defined by a variety of benchmarks. Specifically the ATSDR provided estimates of the prevalence of blood lead concentrations at or above 15 µg/dL, 20 µg/dL and 25 µg/dL, by the use of orthodox statistical techniques.[15]

Because of limitations in the available data, the ATSDR had to confine its attention to people living within what are known as Standard Metropolitan Statistical Areas (or SMSAs). For those SMSAs, figures were available which categorized people according to their age, race, and income.

The ATSDR's central conclusion was that approximately 2.4 million children, that is 17 per cent of children living in those SMSAs, had PbB levels above 15 µg/dL in 1984.[16] The agency also estimated that across the entire US a total of some 3 million children from black and white families would have had blood lead levels elevated above this figure. When other racial groups were included, the ATSDR estimated that between 3 and 4 million children in the entire US might be exposed to sufficient lead to raise their blood lead levels above 15 µg/dL.[17] Although the report does not quite spell it out, that must have meant that at least 5 million had PbB levels above 10 µg/dL.

Since lead can and does cross the placental barrier, any lead which is circulating in the blood of women when they are pregnant will contribute directly to the lead load of a maturing foetus and of newborn infants. The ATSDR therefore also estimated the numbers of pregnant American women, and those of childbearing age, with elevated PbB levels. The ATSDR estimated that about 4 million women of child bearing age, and over 400,000 foetuses were exposed to sufficient lead to raise their blood levels above 10 µg/dL.[18]

Several states now also have a mandatory PbB screening programme for children below six years, by the use of either capillary or venous blood samples. For example between January 1993 and September 1995, 56,379 children were tested in the State of Rhode Island and the results were generally in line with those in NHANES II.[19] The CDC is also collecting data from the National Institute of Occupational Safety and Health (NIOSH) Adult Blood Lead Epidemiology and Surveillance programme (ABLES). They have estimated that in 1994, approximately 42,000 adults were thought to have PbB levels at or above 25 µg/dL.

In 1991 the CDC and the EPA together eventually galvanized both the executive and the legislative branches of the US government to the point where a comprehensive strategy was first developed to address the problem of human exposures to lead pollution.[20] In their 1991 *Strategy for Reducing Lead Exposures*, the EPA estimated that, to a first approximation, approximately a quarter of a million children, or some 15 per cent of all American children below the age of six in the US, had blood lead levels elevated above 10 µg/dL. The EPA also estimated the proportion with PbB levels in excess of 25 µg/dL, and their figures are given in Table 4.1.

Table 4.1 *Estimated percentage of US children under six with elevated blood lead levels in 1990*

Blood lead level (µg/dL)	Percentage above in 1990
> 25	1.0
> 10	15.0

Source: EPA (1991) *Strategy for Reducing Lead Exposures* 21 February, p5

NHANES III
The results of the blood lead component of the NHANES III study were published in 1994.[21] In this project, PbB levels were measured in a sample of 13,201 Americans aged one year or older. During the period from 1988 to 1991 the average PbB level in American children had fallen to just 2.8 µg/dL, as compared with the NHANES II figure of 12.8 µg/dL from the late 1970s; but

> 'Blood lead levels were consistently higher for younger children than for older children, for older adults than for younger adults, for males than for females, for blacks than for whites, and for central-city residents than for non-central city residents. Other correlates of higher blood lead levels included low income, low educational attainment, and residence in the Northeast region of the United States'[22]

Although the percentage of American children aged 1–5 with a PbB level in excess of 10 µg/dL had by then fallen from 88.2 per cent to 8.9 per cent, that still represented approximately 1.7 million children.[23] The percentage of children aged one to two years with PbB levels above 10 µg/dL was even higher at 11.5 per cent.[24]

The only substantial addition to the information generated by NHANES III which has subsequently emerged was published in 1995 by Norman and Clayton Bordley, when they reported on their study of young children in North Carolina.[25] In contrast to the national data, they found that children living in rural areas of North Carolina were at significantly higher risk of lead exposure than urban children. They measured PbB levels for a relatively large sample of 114,034 children below the age of six years, all of whom were screened between 10 January 1992 and 30 September 1994. In total they found that more than 20 per cent of those children had PbB levels above 10 µg/dL. They observed that 'These findings can be explained by the distribution in the state of children living in poverty as well as in pre-1950 housing...'[26] No doubt that is correct, but it does suggest that it might be overly optimistic to suppose that the problem of lead pollution (in either the US or the UK) is confined to urban inner-city areas.

Blood lead levels in the UK

Before the late 1970s, the British authorities made no efforts whatsoever to monitor blood lead levels in the British population. When the Royal Commission on Environmental Pollution reported in 1983, they explained the position unambiguously: 'There are no data on the blood concentration of lead in the UK from which it is possible to generalise about the population as a whole.'[27] The US NHANES II study had started in 1976 and had finished by 1980, but the UK authorities were once again in less of a hurry to deal with the problem of lead pollution than those in the US. Some of the earliest British work did take place in the late 1970s, but British research was far more fragmented and unrepresentative.

In 1977 the British government set about collecting blood and drinking water samples from the homes of some 300 English infants, to study the relationship between lead exposure and PbB levels, but most of the homes selected had low levels of lead in their drinking water, and this made it hard to reach conclusions about the crucial relationship between these two factors, or to generalize to the wider community.[28] The next step was a study of a community with a relatively high level of lead contamination namely Glasgow, Scotland.

The Glasgow study was, like its predecessor, jointly funded by the Department of the Environment (DoE) and the Ministry of Agriculture, Fisheries and Food (MAFF), because they were worried about contamination both in the diet and the water supply, and in other media too. The investigators, and the Departments which provided the funding, knew that Glasgow's drinking water was very acidic and therefore likely to dissolve lead in the plumbing, and that Glasgow's plumbing then contained a great deal of leaded pipes and water tanks. In what came to be known as 'the Glasgow duplicate diet study', the researchers studied pregnant women and their babies, and eventually gathered data on 131 mothers and infants in the period from 1979 to 1980.[29] Mothers were asked to provide investigators with a duplicate of the diet which they had fed to their infants when they were about three months old, and that was analysed for its lead content.[30] Sherlock and Quinn reported that 90 per cent of the infants had PbB values above 10 µg/dL, 36 per cent were above 25 µg/dL, and 13 per cent were above 35 µg/dL.[31] They also reported that the geometric mean PbB level of the mothers was 18 µg/dL, while the corresponding figure for the infants was 17.8 µg/dL. Crucially, however, they pointed out that while the geometric mean for babies who were wholly bottle-fed was 20.8 µg/dL, for breast fed babies it was less than half, at 9.7 µg/dL, a result which provides yet another indication of just how beneficial breast feeding can be.[32]

The first attempts by the British authorities to conduct a survey of blood lead levels which even started to approach a representative sample was a response to a requirement imposed on all Member States of the European Community by a 1977 Directive.[33] This Directive required Member States to conduct two blood lead surveys separated by an interval of at least two years. For each million inhabitants, at least 50 had to be screened, although those occupationally exposed were to be excluded. Since the UK population was then just under 55 million, that corresponded to a minimum sample of slightly fewer than 3000 people. In the event, just under 5,000 samples were collected.[34] The Directive specified, moreover, that if more than 2 per cent of any group had PbB levels above 35 µg/dL, or more than 10 per cent were above 30 µg/dL, or more than 50 per cent were above 20 µg/dL, then action had to be taken to identify the source of their exposure and reduce it.

The first of these two surveys was conducted in the 12 months starting in spring 1979, while the second was conducted in the spring and summer of 1981.[35] A modest amount of information concerning the first study emerged in the 1980 Lawther report, or at any rate in Appendix 2. The investigators reported that PbB figures had been gathered from 2234 adults and 2400 children. Those samples included a randomly selected set of adults living in urban areas, plus several groups thought to be particularly at risk such as the children of lead workers, children living near industrial lead works, and people living near to main roads.[36]

The results of the first of those studies showed that the geometric mean PbB level among adults living in inner city areas was 12.8 µg/dL, while for those in outer city areas it was 11 µg/dL.[37] The city with the highest mean value was Manchester at 17 µg/dL, while the level in inner London was 12 µg/dL. The investigators also reported that 3.1 per cent of the adults surveyed from inner city communities had PbB levels above 25 µg/dL. The results for Manchester indicated that all the adult males had PbB levels above 10 µg/dL, while the same was true for 89 per cent of females. Approximately 45 per cent of adult men in Manchester had PbB levels above 20 µg/dL.[38]

The results for the children included in the first survey found that the geometric mean PbB level in newborn children was 15 µg/dL, and it peaked at the age of two years at a level of approximately 18 µg/dL, declining to a mean of 14 µg/dL by the age of ten.[39] The two surveys did serve to show that the (already rather generous) standards set in the 1977 Directive were being exceeded in several locations in Britain. In particular, exceedances were found in Chester and Leeds, in both cases in the vicinity of lead works, and in Glasgow – a city in which the drinking water was too plumbosolvent.[40] The second survey involved the collection of close to 3500 samples, and it showed in particular that levels of lead in both blood and water in Ayr in Scotland were excessive, and that PbB levels were especially high in the London borough of Islington, and close to a lead smelter in Gravesend in the county of Kent.[41]

When, in 1983, the Royal Commission on Environmental Pollution (RCEP) examined the problems posed by lead pollution in Britain they added to their report (but only in an appendix) a set of estimated lead intakes from a range of sources for some selected groups, and some estimates of consequent PbB levels.[42] There are, however, two remarkable features about this appendix, and the calculations it provides. Plausible estimates of the concentrations of lead in a variety of sources are tabulated, but when it comes to estimating consequent blood lead levels the table is surprisingly unforthcoming.

While estimates of the average blood lead levels in British adults are provided for some groups, the choices made are rather curious in two respects. We are told the estimated average blood lead level for adults who smoke and for those who do not. We are also given estimates for adults exposed to high levels in water, in air and in food. But what is missing is an estimate for the groups at highest risk, namely adults who smoke and are also exposed to high levels of lead from water, air and food, too. The RCEP estimated that an adult with high levels of lead in drinking water would have an average PbB level of 20 µg/dL, an adult with high air lead might have a PbB level of 14 µg/dL, and one with high food lead would reach 15 µg/dL. Simple addition suggests that an adult exposed to high levels by all three routes might on average have a PbB level at or above 50 µg/dL.

The even more important omission is that the RCEP provided no figures whatsoever for the PbB levels of British children. That portion of the table was left vacant. This is an occasion when the EPA's UBK model comes into its own, however. The Royal Commission seem to have assumed, in the Table in their Appendix, that children were exposed to lead levels in water of just 10 µg/L, which is odd because earlier in their report they had shown that figure was too often an underestimate. The appendix also provides no figures for levels of lead in soil and house dust, to which two-year-old children would have been exposed, but in these respects data had been included in the body of the report.[43]

Using the figures which were then available, namely an average figure for lead in house dust (of 561 µg/g) and in soil (at 266 µg/g), together with those provided by the RCEP's appendix for food and air, the EPA's UBK model allows one to readily calculate likely mean blood lead levels which would then have occurred in two-year-old British children. Table 4.2 provides estimates of the mean PbB level, and the percentages with PbB levels above 10 µg/dL and 25 µg/dL, using those averages for dust, soil, food and air, and three separate assumptions about levels in drinking water, namely 10, 50 and 100 µg/L.

The toxicological evidence that PbB levels down to 10 µg/dL were harmful was not as convincing in 1983 as it became in the subsequent ten years, but even so those calculations must have been shocking when they were made (perhaps with rather less precision) back in 1983. No explanation has ever been provided of why those estimates were never published.

There were no scientific or technical reasons why the RCEP could not have filled those gaps in the table, but it is just possible that officials were anxious that the information might anger and frighten the public, if only because it shows that, even by the government's then relatively weak standard, a significant proportion of British children were being exposed to unacceptably high levels of lead. Since the RCEP acknowledged that in some homes levels of lead in water could exceed 300 µg/L, it is obvious that a very serious problem was being under-reported.

Table 4.2 *Estimates of mean PbB level in two-year-old children in 1993, and the percentages above 10 µg/dL and 25 µg/dL, as a function of levels in drinking water*

Level of lead in drinking water (µg/L)	Geometric mean blood lead level (µg/dL)	Percentage above 10 (µg/dL)	Percentage above 25 (µg/dL)
10	12.6	67	7
50	14.7	76	12
100	17.1	84	20

It is just possible that the omission was accidental, but it seems unlikely.

Several other, very localized, studies were conducted in Britain between 1972 and 1984, but as the Royal Commission remarked, 'From the limited data available we do not find a clear and consistent picture of blood lead concentration in the UK'.[44]

A study was conducted in the Scottish city of Dundee in the early 1980s in which blood samples were collected from 1665 mothers and their new-born infants.[45] Dundee is a city in which the levels of lead in drinking water were not especially high, and in this study the investigators found that the geometric mean PbB value for the mothers was just under 6 µg/dL, while the corresponding figure for the newborn babies was approximately 4 µg/dL, but little information is available on the distribution of PbB levels amongst those sampled.[46]

The British government (in the form of the Department of the Environment, or DoE) eventually started to invest in a sustained programme to estimate blood lead levels in an approximately representative sample of the British population in 1984. They did so, not so much as a response to the advice of the Royal Commission, but again because they were obliged to do so under the provisions of a European Directive.[47] The political context in which the study was initiated is also interesting and important. The UK Department of Health and Social Security had published a report of a committee chaired by Professor Lawther in 1980 entitled *Lead and Health*, which provided what the government chose to interpret as a broadly reassuring account of the effects of the prevailing levels of lead on human health in the UK.[48] The Lawther Report, as it became known, provoked a substantial and heated debate.

One of the main critiques of the Report was provided by Professor Derek Bryce-Smith and Dr Robert Stephens of the

University of Birmingham. In their riposte, entitled *Lead or Health*, they argued that the Lawther Report substantially underestimated the contribution then being made to body lead loads by vehicle exhausts. Lawther and his colleagues assumed, in effect, that if the lead emitted from vehicle exhausts was not directly inhaled from the air then it would not be absorbed. Bryce-Smith and Stephens argued that lead from motor vehicle exhausts could in turn contaminate the food supply and be absorbed in food, not just by inhalation.[49] They insisted, therefore, that the Lawther Report seriously under-estimated the contribution which vehicle exhausts were making to body lead loads. They identified numerous other shortcomings in the Report, for example, that most of the figures for airborne lead provided by the Lawther Report were measured at heights ranging from 4 to 25 metres. As Bryce-Smith and Stephens point out 'Such heights may well be relevant to the exposure of giraffes or birds...but are irrelevant to the common case of the urban toddler with his/her nose and mouth close to the exhaust pipe level.'[50]

The DoE's blood lead monitoring programme was conducted in a period when the levels of lead in petrol were being reduced, and while an acrimonious dispute about the extent to which lead in petrol was contributing to body lead loads was underway. It was inevitable, therefore, that the results of the PbB monitoring exercise would be carefully scrutinized by those involved in that dispute. The blood lead monitoring programme covered the period from 1984 to 1987; the level of lead in motor fuel had to be down from 0.4 g/L to 0.15 g/L by the start of 1986, under the provision of EC legislation.

The official accounts of the conduct and interpretation of the results of the programme are not entirely coherent. The decision to initiate the programme study was supposedly linked to the decision to reduce levels of lead in petrol. The programme provided evidence that blood lead levels had declined mainly because of the reduction in the concentrations of airborne lead, but subsequently the investigators tried to cast doubt on the conclusion that the reductions in blood lead levels which they had documented were attributable to reductions in the levels of lead in petrol.

In their earliest report, the investigators disclosed that the study was not being conducted, as the EC Directive strictly required, in a representative sample of the population. Since it was focused primarily on the contribution of lead in motor fuel to body lead loads, Quinn and Delves explain that 'In order to avoid as far as possible any problems of changes in exposure to environmental lead *other than from petrol*, surveys were not planned in areas where it was known that the water was plumbosolvent, or where there were lead/zinc smelters or battery factories.'[51] Indeed they explain that the key portion of their sample lived '...in houses fronting heavily trafficked urban roads.' (emphasis added)[52] In their subsequent reports, however, the investigators failed to draw their readers' attention to the fact that the sample was unrepresentative of the British population as a whole, and under-representative because the groups excluded would have had higher blood lead levels than those included. Secondly, when they eventually reported that, over the period from 1984 to 1987, both air lead concentrations and blood lead levels declined, they endeavoured to maintain, unconvincingly in the face of all their own evidence, that the marked decline in PbB levels had nothing to do with reductions in lead emissions from motor vehicles.[53] It would be difficult to make sense of how some of those reports came to be written without assuming that some political pressure had been exerted on the scientists.

The DoE's blood lead monitoring programme started with a sample of 2000 adults and 1200 children.[54] Eight hundred of the adults lived on the 'heavily trafficked roads', and 300 were occupationally exposed, for example as police officers on traffic duty or as London taxi drivers.[55] There were 400 adult 'controls' comprising 100 individuals living in relatively tranquil parts of the Birmingham area '...where the water supply was known not to be plumbosolvent....and 300 adults in a rural village in south-west England where it was known that air and blood lead concentrations were low.'[56] About 1000 children who went to schools located on heavily trafficked roads were sampled when aged 6–7.

The results for 1984 indicated that for the children, the average blood lead levels were close to 10 µg/dL, which means that about half of the sample had PbB in excess of that figure – at a time when the official British standard at which action was called for was 25 µg/dL and the authorities at the US EPA were already arguing that the US standard should have been lowered to 10 µg/dL.[57] The 1984 results showed that, of the locations studied, the city of Bristol had the highest rates of excessive PbB levels. In 1984, 2 per cent of the children screened and 5 per cent of the adults had PbB levels above 25 µg/dL.[58]

By 1985, it emerged that on average, the PbB levels of adults living on busy roads had declined on average by approximately 1 µg/dL, ie from about 13 µg/dL to 12 µg/dL, and they had declined by a similar amount in 1986.[59] The 1985 results for the sample of children suggested a wide variation, with an average reduction of approximately 1 µg/dL, but the changes varied with the location; in some the fall was as great as 19 per cent while in others an increase of 7 per cent was reported.[60] A particularly detailed account has been provided of a sample of 97 two-year-old children in the city of Birmingham, which showed that only 5 per cent had PbB levels below 6 µg/dL, 5 per cent had levels above 24 µg/dL, while the geometric mean was just under 12 µg/dL.[61]

By 1986, just under 1 per cent of the adults surveyed had PbB levels in excess of 25 µg/dL, but since the geometric mean PbB level amongst the exposed adult men was 10.7 µg/dL and amongst the women was 8.0 µg/dL, this must mean that a substantial proportion of them had PbB levels in excess of 10 µg/dL.[62] Three of the sample of 415 male adults surveyed had PbB levels above 35 µg/dL, despite the fact that they were not occupationally exposed to lead, and lived in areas where the water supply was not plumbosovent. The evidence strongly suggested that they reached such high PbB levels because they had dry-sanded some old leaded paint, providing an indication of how hazardous that common practice can be.[63]

The geometric mean in the 1986 sample of approximately 900 children was 7.7 µg/dL, and while only one child had a PbB above 25 µg/dL, approximately a quarter of them had PbB levels in excess of 10 µg/dL. This represented a modest decline since a quarter of the children sampled in 1984 had PbB levels above 12.5 µg/dL.[64] By the final year, 1987, a quarter of the children sampled had PbB levels in excess of 9 µg/dL, implying that the proportion above 10 µg/dL had by then declined to approximately 20 per cent.[65]

When these results are interpreted, it is important to reiterate that they are not entirely representative of those most at risk, precisely because the data were collected from families living in communities who would have received drinking water containing low levels of lead. It would therefore be over-optimistic to assume that, for example, in 1986 no more than one quarter of all British children had PbB levels above 10 µg/dL.

The main results of the entire study have been represented most clearly by the investigators graphically, using what are called 'box plots'. Their graph is reproduced in Figure 4.3. For each group, five positions are indicated. The central position of the rectangular box is the median value (which is the typical value in the sense that it is the midpoint of the observations when they are arranged in ascending order), while the lowest point on the line is that attained by just 5 per cent of the sample. The point at which the lower thin line reaches the bottom of the rectangular box indicates the PbB level reached by 25 per cent of the sample, while the point at which the upper thin line joins the top of the rectangular box corresponds to the level reached by 75 per cent of the sample, and the very top of the thin line

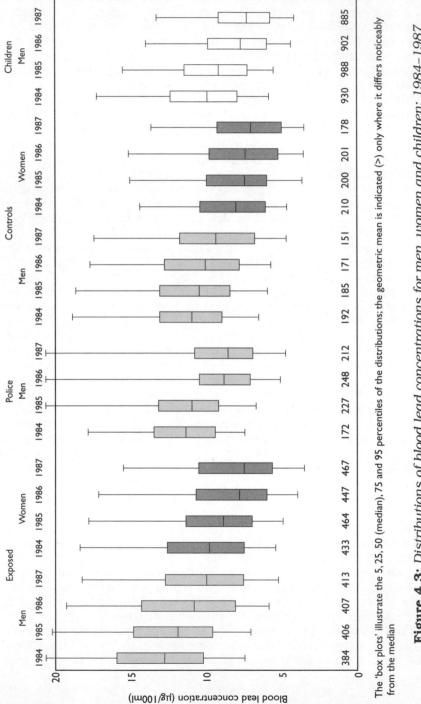

Figure 4.3: *Distributions of blood lead concentrations for men, women and children: 1984–1987*

The 'box plots' illustrate the 5, 25, 50 (median), 75 and 95 percentiles of the distributions; the geometric mean is indicated (>) only where it differs noticeably from the median

represents the level reached by just 5 per cent of the sample.

The data represented in Figure 4.3 indicate that all groups were then experiencing a decline in their average PbB levels, and that levels were generally declining by about 4–5 per cent a year.[66] The decline was independent of factors such as age, social class, drinking habit or the age of buildings in which the subjects lived. As indicated before, however, the investigators were unwilling to attribute these reductions directly to the reductions in lead emissions from motor vehicles. Emissions of lead from motor vehicles declined, for example, by 50 per cent in urban areas between 1985 and 1986.[67] Nonetheless, the authors argued that the fall in PbB levels could have been due, in part, to reductions in exposure to lead from other sources such as water and food which occurred over the same period.[68] This suggestion was and is entirely unconvincing for several reasons. Firstly, as has already been explained, the communities sampled received low levels of lead in their drinking water, and therefore improved water treatment to reduce plumbosolvency would have had no impact on these people. Secondly, the Ministry of Agriculture, Fisheries and Food reported elsewhere that although the levels of lead in food had declined over that period, that decline would not have been sufficient to account for the reductions in PbB levels found by the team working for the Department of the Environment. Thirdly, in so far as reductions in the levels of lead contamination in food did contribute to reduced PbB levels, that in turn would have been partly a consequence of reduced emissions of lead from motor vehicles. It is hard to avoid the conclusion, therefore, that Bryce-Smith and Stephens had been right all along, and that Lawther and his committee had been mistaken about the contribution of leaded motor fuel to body lead loads, and that the UK government had been unwilling to accept the evidence which they had themselves produced.

It seems particularly unfortunate, moreover, that the Department of the Environment's study was not extended beyond 1987, because the blood lead monitoring ceased at the same time as the government introduced a lower rate of tax on unleaded petrol, which in turn dramatically boosted sales.[69] A time-series data set was therefore interrupted at precisely the crucial moment when it might have been possible to resolve the long-standing dispute about the proportion of PbB which could be attributed to vehicle exhausts or to explore the impact of the growing use of unleaded petrol on PbB levels. Representations were made to the Department of the Environment recommending extending the UK Blood Lead Monitoring programme beyond 1987, but the Department steadfastly refused to provide the resources.

The discussion above has provided a review of the published information. It is clear that several other studies have taken place, but for reasons which remain difficult to discern, decisions have been taken not to publish the results. For example, between 1984 and 1992, about 35,000 children in the Leeds-Bradford conurbation were screened in a programme run jointly by the local Health Authorities and the Environmental Health Departments of the city councils, but the results remain unpublished. My understanding, however, is that just under 2 per cent of those children (corresponding to approximately 425 children) had PbB levels above 10 µg/dL, and about 0.5 per cent (ie approximately 125 children) had PbB levels above 25 µg/dL.[70]

In 1995 the Department of the Environment and the Department of Health collaborated in a small study, which, as it were, piggy-backed on the Health Survey of England, which was intended particularly to examine the relationship between blood lead levels and blood pressure levels in adults not occupationally exposed to lead. This investigation involved scrutinizing PbB levels for 6868 people, 95 per cent of whom were over the age of 16, between January 1995 and February 1996 when their blood was sampled. One hundred and eighty boys and 160 girls aged 11 to 15 were also included, but younger children were not screened. PbB levels above 10 µg/dL were found in 5.3 per cent of the men and 1.2 per cent of the women, while all of the 11–15 year olds had PbB levels below this figure.[71] The data from this recent study suggest that average adult PbB levels declined in the period from 1987 to 1996, but they do not provide any information on the most vulnerable group, namely those aged six and under. These latest British results suggest that average PbB levels have been falling quite rapidly, but they provide no information about the numbers of young children with PbB levels above 10 µg/dL, and there are good reasons for supposing that these have not been falling so rapidly. The reduction in the quantities of lead emitted by motor vehicles must be welcomed, but the problems caused by old paint and water pipes remain to be addressed.

A small but interesting study was conducted in 1994 in the Lancashire town of Blackburn, but the results have never been published. I have, however, seen a copy of the results. The initiative for the work came from the local council and local health department, but much of the funding came from the Department of the Environment. The Blackburn study found high levels of lead in water, one of which was as high as 88,200 µg/L. A major limitation with this study, however, is that there is no way of telling the extent to which the sample screened was representative of the population as a whole. Investigations in particular households were initiated only when a consumer complained about receiving cloudy tap water. Cloudiness is, however, largely irrelevant to levels of lead pollution; water that does not appear cloudy may nonetheless be highly contaminated.

The number of people willing to provide blood samples, in this study, was very low. Only 24 subjects from ten households provided blood samples. Five of the samples taken from just eight children had PbB levels above 10 µg/dL, and seven of the samples taken from 16 adults exceeded this figure.[72] The highest child PbB figure was about 25 µg/dL, while the highest adult figure was about 35 µg/dL. Some of the scientists participating in this study examined the isotopic character of the lead found in the drinking water, and they found that the lead in the water originated predominantly from the water pipes, rather than from any prior source of contamination. Since the water supplied to Blackburn has long been treated to reduce plumbo-solvency the evidence from Blackburn demonstrates that chemically reducing acidity will not by itself solve the problem of excessive lead contamination.

The Avon Longitudinal Study of Pregnancy and Childhood (ALSPAC) is based at the University of Bristol, and it involves a long-term scrutiny of a representative sample of mothers and their children drawn from the population of the English County of Avon. The total sample includes 14,000 children born since 1990. A subgroup of that population, consisting of 1135 children, were invited in 1995 to participate in a detailed study of their development at age 31 months. The investigators were able to analyse blood samples taken from a total of 584 of those children. Only two of those children had PbB levels above 25 µg/dL, but just over 30 of the children (ie 5.4 per cent) had PbB levels above 10 µg/dL. The geometric mean PbB levels amongst both the boys and the girls was approximately 3.4 µg/dL. Children whose parents smoked, and those with pets in the home, or those who lived close to busy roads, were the ones with higher PbB levels. Indeed, the investigators have analysed the PbB data by reference to the postcode (the British equivalent of the zip code) and found that the main problems were confined to a handful of postal districts, namely (in descending order of severity) BS4, BS5, BS3, BS7 and BS11.[73]

As far as it is possible to tell, the available evidence suggests that somewhere between 5 and 10 per cent of the British population have PbB levels in excess of 10 µg/dL. Moreover, it is generally understood and accepted that young children, particularly those who live in pre–1945 homes, and who receive their drinking water through lead pipes, will be amongst those with the highest PbB levels, except that is for adults working in and around lead smelters and battery plants. The first aim of public policy in this regard therefore should be to identify those children and to improve their environments. If there is anyone who knows how many young British children have elevated blood lead levels, then they are not yet prepared publicly to acknowledge it.

Acute lead poisoning in the UK

There are many serious shortcomings in the way in which the UK deals with the most serious aspects of lead exposure, when compared with the US. A particularly important example concerns cases of severe acute poisoning.[74] At present, national figures on the incidence of acute lead poisoning in the UK are neither being collected nor reported.[75] Britain does have a National Poisons Unit, but it is unable to estimate the number of people who are suffering from acute lead intoxication.[76] Acute lead poisoning, in so far as it is being diagnosed at all in the UK, is being detected on an unsystematic basis which is dependent on individual medical practitioners and their diagnostic skills. The British government has set a maximum PbB target of 25 µg/dL for those not occupationally exposed to lead, but there is no consensus amongst clinicians in the UK on the PbB level that constitutes lead intoxication and inconsistent criteria are being used.[77] This contrasts sharply with the US where, as explained previously, several states already have mandatory PbB screening programmes for children below six years, and where the Adult Blood Lead Epidemiology and Surveillance program is operating.[78]

Blood lead levels in six-year-old children in the UK

Despite all these shortcomings, some limited progress can be made in the UK by utilizing the computer model of the uptake and biokinetics of lead in the human body which has been developed by the US EPA.[79] As explained above, the model allows users to estimate probable patterns of the distribution of blood lead levels in a population of children by inference from estimates of the levels of lead contamination in specific parts of the environment, such as air, food, drinking water, soil and dust. The calculations provided in Table 4.3 have been made using some of the estimates which are available on the levels occurring in the British environment. The environmental contamination data

94

come from the work of Thornton et al, who have provided the most comprehensive and up-to-date estimates of levels of lead in the British environment available.[80] Columns 2 and 3 of Table 4.3 show some of their data. The results obtained by introducing those data into the EPA's model are given in the final two columns and indicate the estimated percentage of children under the age of six, in each of several areas, who can be expected to have PbB levels above 10 and 25 µg/dL.

These estimates imply that in the early 1980s, with the exception of old lead mining sites, a substantial majority of British children aged six years or less had PbB levels below 25 µg/dL, but that a significant proportion had PbB levels above 10 µg/dL, especially those living in homes built before 1960. We are entitled to assume, by reference to the considerations reviewed in Chapter 3, that their mental development suffered some consequent impairment. The problem with the figures in Table 4.3 is that they are based upon the only data available, and they are very out of date. This suggests, however, that for a relatively modest investment in environmental lead monitoring, and without having to take blood samples from young children, useful estimates could be made of the lead loads which children are currently carrying.

Lead contamination of the environment in the US and the UK

The levels of lead which occur naturally in both surface and ground water sources are generally very low. Lead concentrations in natural water sources vary according to the chemical characteristics of the rock substrate from which they derive. They are sufficiently low that naturally occurring lead is hardly ever a significant source of human exposure. Most of the lead which is found in drinking water derives from human interventions.

Almost all of the lead contamination of the drinking water supply occurs in the distribution system, arising from lead pipes and lead-bearing materials such as solder and alloys used in plumbing fittings such as taps or faucets. In some areas of both the US and the UK there are substantial numbers of homes where water is still delivered through old lead pipes, and it is even occasionally stored in lead-lined water tanks. There is also evidence indicating that when pipework has been newly installed, it is especially easy for the lead in lead-based solders to leach into the water, but as it ages, the rate at which leaching occurs declines.[81]

Table 4.3 Levels of lead in house dust and soil in the UK in 1982# and estimated rates of elevated blood lead (PbB) levels in children under six years of age

Location	House dust μg/g*	Soil μg/g*	Estimated percentage of <6-year-old children with PbB >10μg/dL	Estimated percentage of <6-year-old children with PbB >25μg/dL
Mean value in national survey** (range)	561 (5–36,900)	266 (13–14,100)	26.9	0.5
London (range)	1010 (5–36,900)	654 (60–13,700)	55.5	3.9
Birmingham	424	313	21.1	0.3
Derbyshire old mining villages (range)	1870 (606–7,020)	5610 (1,180–22,100)	97	50.1
Brighton houses built 1870–1919	1874	1014	81.1	14.5
Brighton houses built 1960–1986	241	131	8.3	<0.05

Measurements in Brighton were taken in September 1986; * Geometric mean of multiple measurements;
** Including London, but excluding old mining areas

Source: Thornton, I et al (1990) 'Lead Exposure in Young Children from Dust and Soil in the UK' *Environmental Health Perspectives* Vol 89, pp55–60

Dietary Pb intake set at 15.71 μg/day on the basis of figures provided by MAFF (1989) *Lead in Food: Progress Report*, Food Surveillance Paper No 27, HMSO, pp13–15

Conservatively assuming water lead concentrations are, on average, 4 μg/litre; and ambient air lead concentrations

The extent to which drinking water contains lead, however, depends on a number of other factors. First and foremost, the acidity of the water is important in determining the degree to which lead is leached from the plumbing into the water. Lead is relatively insoluble in alkaline water. In areas with chalky soil and water, the water pipes are often lined with relatively insoluble deposits of lead carbonate and calcium carbonate. These areas usually have relatively low lead levels in their drinking water, but it is still possible for significant contamination to occur. If the piping is badly deteriorated or subject to vibrations, for example from heavy traffic, especially buses and trucks, fragments of the lead salts may become dislodged and crumble into the water. When the water containing those particles is drunk, the acid in the human digestive system will readily dissolve the lead, and it thereby becomes what toxicologists call 'bio-available'. In acidic water, however, the problem is generally worse.

The temperature of the water is also important in determining the quantity of lead in the water, since lead and its compounds are more soluble in warm water than in cold. A hot water system with lead soldered joints can contain far more lead than a cold water system.[82] If a family were regularly to fill its kettle from the hot water system, this could substantially increase their lead exposure.

The level of lead in the water emerging from a tap or faucet is also dependent on the length of time for which the water has been stationary in the pipework. Typically, lead concentrations are at their highest in the very first flush, for example when left overnight, and as the system is flushed, concentrations decline; the time taken for lead levels to reach their equilibrium levels therefore depends on the length of lead piping and the rate of flow. These considerations influence precisely how one specifies the standards for measurement of levels of lead in water, and the form in which regulations may be cast. In many households, but not all, the first flush water is routinely consumed.

Lead in drinking water in the US

Lead contamination of drinking water can occur under a number of conditions. Older properties may have leaded water pipes – an obvious source of contamination. On the other hand, newly soldered pipework, on which leaded solder has been used, also poses a hazard. The lead from the solder can be readily leached out by acidic water, and the resultant levels of lead contamination are particularly high during the first couple of years after the solder has been used. Just about all the lead that can be leached out of the solder will have been leached within an average period of about five years.[83] The more acidic the water is, of course, the more likely it is to leach lead from the pipework. The use of lead water pipes in American homes was particularly common before about 1920, after which iron pipes became increasingly popular, while the use of lead solder was very common until the early 1990s.[84]

A large-scale national survey of lead in drinking water, funded by the US EPA, took place in the US in 1981. In that study Patterson tested samples collected by the Culligan Water-Softening Company, but the water samples were not taken from either the first flush nor as fully flushed-out samples, but as mid-stream specimens, after the water had been flowing at a moderate pace for 30 seconds.[85] The resultant estimates of the percentage of samples with lead concentrations at various levels are given in Table 4.4.

Table 4.4 *Estimated percentage of drinking water samples collected in 1981 in the USA with various lead concentrations*

	Measured lead concentration (µg/L)			
	≤10	11–19	20–49	≥50
Percentage of samples	60	24	13	3

Source: EPA (1986) *Reducing Lead in Drinking Water: A Benefit Analysis*, p11–22

These figures are, inevitably, imprecise and should be interpreted as order of magnitude estimates, but they do indicate that while the majority of partly-flushed samples were below 10 µg/L, a significant minority showed elevated lead levels.[86]

In an attempt to estimate the degree to which children were exposed to lead from drinking water in the US, the ATSDR produced calculations for three groups within the population. Firstly, those which were defined as 'potentially exposed', secondly those with actual measurable exposures at levels which were not necessarily toxic, and thirdly those exposed to toxicologically significant levels.[87] A child was deemed 'potentially exposed' if it lived in the kinds of homes built at times and in areas in which lead piping was commonly used. The US Census Bureau had estimated that in the early 1980s, approximately 5.2 million children under five years old and 8.7 million children from 5 to 13 years lived in such properties, and an undetermined proportion of those buildings contained old lead service connections.[88]

The Department of Housing and Urban Development provided some more precise information concerning the age distribution of the American housing stock. On the basis of those figures, the ATSDR estimated that 'In 1983, of the 21 million US children under 6 years of age, 13 per cent or 2.73 million lived in units that had lead water service connections. Similarly, 4 per cent or 840,000 children lived in homes built within the last 2 years, which is the housing fraction having lead-soldered new copper plumbing.'[89] The ATSDR also provided estimates of the number of people receiving water that was sufficiently acidic and therefore corrosive to leach lead from old pipes and solder. The EPA used those figure to estimate that in 1986 approximately 62 million people in the US received such acidic water, and of them some 11 per cent or 6.8 million were children under the age of seven years.[90]

It was also possible to estimate the numbers of children exposed to drinking water containing sufficiently high lead levels to elevate their blood lead levels. To do this, the ATSDR adopted the EPA's assumption that if a child is exposed to water containing more than 20 µg/L of lead, its blood lead level would consequently be elevated.[91] As many as 42 million people were estimated to be receiving drinking water containing more than 20 µg of lead per litre in 1988.[92] The agency then calculated that roughly 3,780,000 children under six years of age (or 9 per cent of the population) are likely to have been consuming such water.[93]

Since the late 1980s, when the ATSDR's report was published, steps have been taken in some communities both to reduce the acidity of the drinking water and to replace lead service lines, and consequently average levels of lead in drinking water have fallen. Detailed up-to-date figures are available for some localities, but the nation-wide picture has not yet been fully documented. Current thinking in the EPA is that in the US, on average, less than 10 per cent of a child's typical intake of lead comes from drinking water, but the problem has not yet been dealt with in all neighbourhoods, and for the children who live in those neighbourhoods, drinking water is contributing substantially to body lead loads.

In June 1991, the EPA issued a Lead and Copper Rule which initiated the most thorough national survey of lead in drinking water ever in the US. By 1993 the EPA had gathered the results for the second round of monitoring for large systems (defined as a single distribution system serving more than 50,000 people) and the first round for medium-sized systems (ie those serving between 3300 and 50,000 people). The EPA's rule also required the operators of all large systems, and those smaller systems which exceed the Agency's lead action level, to take steps to reduce consumers' risk of exposure to lead.

In 1992, the operators of 682 large water systems were required to conduct lead monitoring between January and June 1992 at 'high-risk' residences (ie those containing lead-based plumbing), and again at the same residences between July and December of 1992. There were also approximately 7000 medium-sized systems, for which monitoring had to be conducted between July and December of 1992.[94] Those instructions were not, however, fully implemented. The EPA reported that in 1993, 1100 large and medium-sized systems had failed to complete their monitoring, and consequently the EPA and the states have between them issued some 427 notices of violation and 145 proposed or final administrative orders to public water systems for their failure to monitor.

One hundred of the large systems, serving an estimated 21 million people, and 719 medium-sized ones, serving approximately 9 million people have, moreover, reported lead levels in excess of the EPA's 'action level' of 15 µg/L (ie 15 parts per billion) for more than 10 per cent of the residences served. An 'exceedance', as it is termed, is not by itself a violation of the law, but it entails that certain actions should be taken. Those actions should include, for example, additional monitoring, public education and corrosion control. In communities where the action level is regularly exceeded, the EPA advises residents to let the water flow for at least 15–30 seconds before drinking it or using it for cooking, and not to use the hot water supply for cooking or for preparing infant and baby formulas.

The highest lead level found anywhere in those parts of the water supply system which were tested was 484 µg/L, which is 32 times higher than the EPA's action level. It was found at the US Marine Corp's Camp Lejeune at Hadnot Point, North Carolina. The exceedances for medium-sized systems were greater than for the large ones, which indicates that the EPA is acting prudently when it insists on not confining their attention to the largest systems, despite proposals to that effect from some quarters.

Unless and until all water suppliers have conducted the requisite monitoring, and their results have been reported, it is not possible to provide an up-to-date assessment of the position, but the EPA is endeavouring to ensure that wherever excessive levels of lead are detected and reported, action is then taken to diminish the risk of exposure.

Lead in paint and dust in the US

Large amounts of lead, in some cases even up to 50 per cent of total dry weight, were contained in domestic interior and exterior paints used in the US before the 1950s.[95] According to the Department of Housing and Urban Development (HUD),

> *Lead-based paints have been produced since ancient times. The first factory to produce white-lead pigments in the United States was established in 1804 in Philadelphia. Paints with lead-based pigments were highly regarded for their durability, adhesion and hiding qualities. Based on the history of the production of white-lead pigments relative to other pigments, lead concentrations in paint manufactured in the United States were probably highest during the first two or three decades of the 20th century. However, lead-based paint remained in widespread use during the 1930s and 1940s and to a declining extent into the 1970s.[96]*

The presence of lead was even thought of as a key selling point in the 1930s.[97]

The Department of Housing and Urban Development estimated in 1990 that 14 million housing units contained unsound or deteriorated leaded paint and that 3.8 million of these households were occupied by young children.[98] Across the entire nation, the EPA estimated in 1991 that about 3 million tons of lead was contained in the paint of roughly 57 million occupied private homes that were built before 1980.[99]

This is of particular concern because it is widely agreed that old leaded paint constitutes the major source of lead exposure, and old peeling and flaking leaded paints may give rise to relatively high risks of both chronic and acute lead poisoning. Children are particularly at risk of elevated PbB levels from this source because they engage in more hand-to-mouth activity than adults and can therefore ingest significant quantities of contaminated dust. Moreover, young children often engage in pica (ie the habit of putting inappropriate items in the mouth, which may be contaminated with lead).[100]

The population exposed in the US

The ATSDR has used several methods to provide estimates of the numbers of children exposed to leaded paint. Firstly, they

Figure 4.4 *Advertisement for 'Dutch Boy' paint, 1937*

used data from the US Bureau of the Census 1983 survey to estimate the total number of children living in housing decorated with old leaded paint, which was defined as paint containing no less than 0.7 mg of lead per cm^2. They estimated that there were almost 42 million lead-painted residential households in the US in 1988, which then corresponded to over 52 per cent of the national housing stock.[101] Given the turnover in the US housing stock, it is reasonable to assume that the proportion had by the mid-1990s declined to just below half. The ATSDR went on to estimate that roughly 12 million children under seven years of age lived in housing, regardless of age or state of repair, containing leaded paint.[102] They also estimated that about 80 per cent of those children lived in urban areas.[103]

The ATSDR's next step was to estimate the number of children in the US living in deteriorated housing containing leaded paint. By reference to the work of Pope, and data from the 1983 Housing Survey, the agency concluded that approximately 1,772,000, and probably no more than 1,996,000, children were living in run-down housing.[104]

When the Department of Housing and Urban Development published its plan for lead paint abatement in 1990, information was provided which supplemented the estimates given by the ATSDR.[105] HUD indicated, in particular, that in 1989 and 1990

Of the 77 million privately owned and occupied homes built before 1980, 57 million, or three-fourths, contain lead-based paint. Of these 57 million units, an estimated 9.9 million are occupied by families with children under the age of 7, who are most at risk from lead poisoning. However, a much smaller number of units have conditions that pose priority hazards: 3.8 million of the units have peeling paint, excessive amounts of dust containing lead, or both problems. Of these, 1.8 million are occupied by children whose families have incomes above $30,000, which is approximately the median income for all households; 2.0 million are occupied by lower-income families with children, of whom 0.7 million are owner-occupants, and 1.3 million are renters.[106]

Lead in dust and soil in the US

The majority of the lead which is found in dust and soil derives from sources such as paint, motor vehicle exhausts and deposition from stationary sources such as industrial installations. It is therefore difficult to separate the contributions to overall exposure of lead-contaminated soil and dust from those other sources. The relative importance of dust and soil as sources of exposure is, moreover, likely to vary enormously, depending on a multitude of complex factors. For example, the weather may influence the amount of time that young children spend out of doors and potentially exposed to lead contaminated soil, or remain indoors and exposed to domestic dust.[107]

Levels of lead in dust and soil are generally found to be a function of the locally prevailing levels of lead in paint, vehicle exhausts and stationary industrial sources.[108] While levels found in some rural areas may be vanishingly slightly, in some urban areas they range well above 1000 µg/g, but as the ATSDR has pointed out, there are no standardized methods for collecting samples, and therefore it is particularly difficult to generalize from the results which have, so far, been published.[109] The conclusion reached in 1988 by the ATSDR, however, was that between 5.9 and 11.7 million American children were potentially exposed to significant levels of lead from dust and soil.[110]

The EPA responded to that evidence in a thorough fashion, and initiated a detailed project which came to be known as the Three Cities Lead Soil Abatement. After considerable time and effort, however, the EPA, perhaps somewhat reluctantly, concluded that '...lead-contaminated soil abatement is not likely to be a useful clinical intervention for the majority of urban children in the United States with low-level lead exposure.'[111] They found, in short, that soil abatement was far less helpful for reducing childhood PbB levels than the abatement of leaded paint and drinking water.[112]

Lead in air in the US

Leaded gasoline
Lead additives were introduced to improve the performance of the fuel, by controlling the combustion process. Millions of tons of lead from the combustion of leaded gasoline have been dissipated into the environment of the US since lead additives were first used in the 1930s. Lead levels in the air, and fallout derived from automobile exhausts, were typically found at their highest in urban and suburban areas, and more particularly close to heavy road traffic. Surveys by the National Filter Analysis Network, for example, have shown that while ambient air lead levels in the late 1980s in remote parts of the US were around 2 ng/m^3, those in urban areas were some 1000 times higher, in the range 1–3 µg/m^3.[113] The historical pattern of the use of lead in American gasoline is shown, from 1930 to 1992, in Figure 4.5.

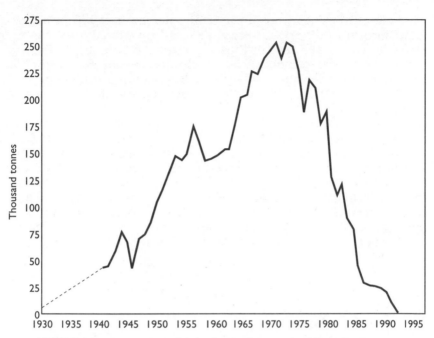

---- 1930–1941: Inferred usage – data included under miscellaneous and undefined categories
Source: OECD (1992) p95; derived in turn from US Bureau of Mines, 1992

Figure 4.5: *Lead used in the production of gasoline
in the United States: 1930–1992*

This welcome decline in the use of lead in motor fuels has
resulted in marked reductions in the quantities of lead in the air
and consequent reductions in body lead loads. Despite these
reductions, much of this historic lead load remains in the
environment, mainly in dust and soil, and so can contribute to
current exposure.

Stationary industrial sources of lead
Industries such as primary and secondary smelters, battery
plants, incinerators and coal- and oil-fired power stations
produce lead-containing emissions. In the past in particular,
these emissions resulted in substantially raised air lead levels
over considerable areas surrounding the source. For example,
levels of airborne lead near smelters have historically reached
levels between 5 and 15 µg/m^3 at distances of between 5 and 15
km from the site.[114] Controls which have been introduced during
the last 20 years have substantially reduced the quantities of
lead being released into the atmosphere from such installations.

Lead emissions from industrial sources in 1990, such as smelters and refineries, were estimated by the EPA as having been below 50 per cent of those reported in the late 1970s.[115] Data derived from monitoring conducted by the EPA indicate that emissions from stationary sources declined markedly over the period 1980 to 1990, and these are represented graphically in Figure 4.6.

Despite those trends, some contamination of ambient air continues to occur around some stationary sources.[117] The EPA reported that, in 1990, nine areas in the vicinity of non-ferrous smelters, or other point sources of lead, exceeded the National Ambient Air Quality Standard (NAAQS) of 1.5 µg/m^3.[118] Moreover, in 1990 5.3 million people were estimated to be living in counties which did not comply with the NAAQS for lead.[119] A graphical representation of the areas in which the NAAQS for lead was exceeded in 1990 is given in Figure 4.7.

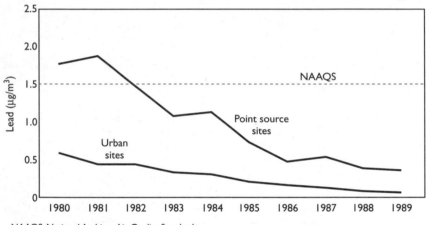

NAAQS: National Ambient Air Quality Standard
Source: EPA (1991) *National Air Quality and Emission Trends Report, 1990* pp3–30[116]

Figure 4.6: *Concentration of lead in the air in the United States: 1980–1989*

Lead in food in the US

Since everyone eats food, dietary lead is a general source of exposure for the entire population. There are, however, great difficulties estimating the contribution which foods make to body lead loads. Firstly, different items of food, such as canned food, shellfish or milk and crops, can and do have very different levels of lead. In addition, people's diets vary enormously. The problem of estimating the contribution to PbB levels from food

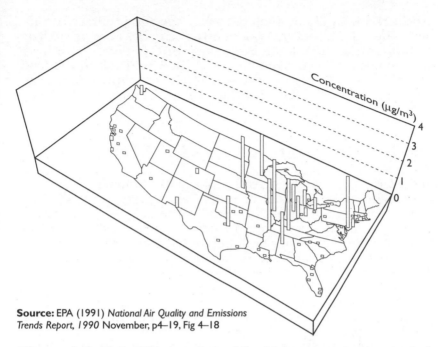

Source: EPA (1991) *National Air Quality and Emissions Trends Report, 1990* November, p4–19, Fig 4–18

Figure 4.7: *United States map of the highest maximum quarterly average lead concentration by MSA, 1990*

is complicated further by the fact that humans absorb lead from the intestinal tract at varying rates, which depend on factors such as age, nutritional status, their prevailing lead body load, and the form in which the lead is present.

Lead can enter the food chain in several ways. For example, it may be absorbed by plants through their root systems and by deposition on their leaves, and in meat it may be a consequence of its presence as a contaminant in animal feedstuffs. Lead contamination of food can occur both in production and during processing. Most dietary lead, however, comes from lead in canned food. Dietary lead can account for a significant proportion of an individual's total body lead load, particularly if they are not exposed to other significant sources.[120]

The ATSDR estimated that in 1988 no more than 5 per cent of children in the US under six years of age were receiving sufficient lead in their diets to contribute more than 10 µg/dL to their PbB levels. In 1988, that corresponded to approximately 973,700 children.

As far as the US is concerned, there is something quite curious about lead in food, because it is the one route of exposure which is not subject to any specific or direct regulation. There is, however, evidence that the US Food and Drug Administration (FDA) has been busy in this area; there have been extensive discussions between the US FDA and the food industry on this topic, and the levels of lead in American foodstuffs have diminished without the FDA having introduced a formal regulatory apparatus.

Lead in drinking water in the UK

In the UK there is a widespread problem of elevated lead concentrations in drinking water; the older the property the more likely it is to have high levels of lead in the tap water. A study conducted in the 1980s found that 15–20 per cent of samples taken from British homes built before 1944 were above 50 µg/L, while only about 5 per cent of houses built after the end of the Second World War exceeded that figure.[121] In 1982 the Berkshire-based Water Research Centre (WRC) told the British government that 'The problem of lead in drinking water is not unique to the UK but there are few other countries where the problems appear to be as severe.'[122]

The public water supply to Glasgow, Scotland, provides a telling example. It was first installed in the 1854, after lengthy consideration and debate. Ten years before the work started, a gentleman named Christison had recommended in the *Transactions of the Royal Society of Scotland* that lead should not be used in the system of water pipes or tanks because of its already-recognized toxicity.[123] His advice was not heeded, and the same mistake was made again when Edinburgh's fresh-water supply system was installed. Consequently, when the levels of lead in tap water in the UK were surveyed in the mid-1970s, 29 per cent of households in Scotland had first-flush tap water containing more than 100 µg/L, while in Britain as a whole the corresponding figure was 9 per cent.[124] The problems in Glasgow were especially severe; in 1979–80 approximately 42 per cent of the water samples had lead concentrations in excess of 100 µg/L.[125] The problem in Scotland was aggravated not just by the acidity of much of the water, but also by the fact that many homes contained lead-lined water tanks.[126]

As long ago as 1977 a group of British researchers showed, by studying samples of the Scottish population, that there was a close relationship between levels of lead in drinking water and blood lead levels.[127] It was, however, not until the mid-1980s that action to diminish this route of exposure was taken. By 1982 the Ministry of Agriculture, Fisheries and Food acknowledged that in the cooking process, lead can be absorbed by food from the water in which it is cooked.[128]

Scientists in the Department of the Environment realized as long ago as the 1970s that there was a problem with lead in drinking water in all too many parts of the UK, and they at least had sufficient support from their ministers to be able to obtain the funding necessary for a reasonably thorough survey of lead concentrations in drinking water which was conducted during 1975 and 1976.[129] The main results of that survey emerged in the Lawther Report, and are reproduced as Table 4.5.

Table 4.5 *Lead concentrations in tap water in Great Britain in the late 1970s*

| Lead concentration (µg/L) | Percentage of households in | | | |
	England	Scotland	Wales	GB total
0 – 9	66.0	46.4	70.5	64.5
10 – 50	26.2	19.2	20.7	25.3
51 – 100	5.2	13.4	6.5	6.0
101 – 300	2.2	16.0	1.5	3.4
301 and above	0.4	5.0	0.8	0.9

Source: DHSS (1980) *Lead and Health* p18, Table 14

The figures indicate that, averaging over the whole of Great Britain, approximately 35 per cent of random daytime drinking water samples contained lead concentrations greater than 10 µg/L. Twenty-five per cent of samples fell within the range from 10 to 50 µg/L. In Scotland, as many as 5 per cent of all drinking water samples were found to be above 300 µg/L, while more than half of all Scottish samples had lead concentrations of above 10 µg/L, though in England and Wales the corresponding figures were below 1 per cent.[130] A study of lead levels in drinking water in Northern Ireland conducted during the late 1970s found that 3 per cent of random daytime samples were above 50 µg/L, and 1 per cent were above 100 µg/L.[131]

The Water Research Centre estimated in 1982 that somewhere between 7 and 10.5 million, out of a total of about 18.5 million households in the UK, had lead piping somewhere in the connection between the water main and the kitchen tap.[132] That corresponded to somewhere between 38 and 57 per cent of all households. They estimated, moreover, that approximately 5 million of those households, ie 27 per cent of them, were then being served by water which was sufficiently acidic to be plumbosolvent.[133] The WRC then estimated that the average cost of replacing the lead pipes was £587 per household, of which £203 was for the water company's communication pipe, and £384 was to cover the householder's pipework.[134]

In 1981, a group of leading British investigators examined the effects, in Ayr in south west Scotland, of reducing the acidity of the drinking water on consumers' blood lead levels.[135] The town was chosen because the second of the two DoE surveys conducted in the late 1970s had found disturbingly high levels of lead in the drinking water.[136] Starting in 1981, a year after the EEC Directive had set a maximum at 50 µg/L, considerable quantities of phosphates began to be added to the water supply. Prior to phosphate treatment, that standard was not being met. Only 15 out of 66 water samples taken in 1980 (prior to dosing with phosphate) had lead levels below 50 µg/L, 20 samples were below 100 µg/L, and 26 water samples were above 300 µg/L.[137]

The acidity or alkalinity of water is measured in terms of its pH. A neutral solution is defined as having a pH of 7, while higher figures represent increasing alkalinity, and lower numbers connote acidity. Phosphate was added, in this study, in sufficiently large concentrations to shift the pH from the acidic range between 4.5 and 5.5 up to the alkaline level of 8.5. Once phosphate dosing was introduced, the next set of water samples all fell below 300 µg/L, and 25 samples out of 66 were below 50 µg/L.[138] Sherlock et al reported the not particularly surprising result that water treatment produced a sharp fall in water lead concentrations, and a consequent decrease in the median blood lead concentration from 21 to 13 µg/dL.[139] The investigators did an important service by taking the trouble to analyse their data in an illuminating fashion. The statistics were quite complicated, but the underlying concept is straightforward. They investigated what toxicologists like to call 'the shape of the dose–response relationship'. It can be understood most easily in graphic terms, and their key result is shown in Figure 4.8.

Their analysis is clear evidence for the crucial conclusion that '...successive decreases in water lead concentration will yield progressively larger decreases in blood lead concentrations, and that even relatively low concentrations in water will have a marked effect on the concentration of lead in bloods.'[140] This is important because it implies not only that reducing the water lead limit from 50 µg/L to 10 µg/L provides significant benefits, but also that diminishing it significantly further would confer real benefits.

In the mid-1980s a team headed by two local government environmental health officers conducted an interesting and

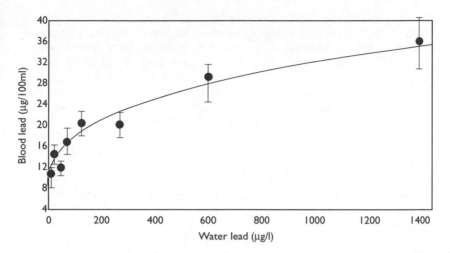

Figure 4.8: *The relationship between blood lead concentrations and water lead concentrations in the Ayr study*

Source: J C Sherlock et al (1984) 'Reduction in exposure to lead from drinking water and its effect on blood lead concentrations' *Human Toxicology* Figure 2, p390

detailed study of the levels of lead in the drinking water of households in a middle-class suburb of Glasgow called Renfrew.[141] Every house within the district that was suspected of containing lead pipes was surveyed. The local council was informed by the water company that lead had not been used in the water pipes in that area 'after 1968'. (The contrast with the US is striking. Americans had stopped using lead for water pipes in the 1920s, while the practice in Scotland came to an end 114 years after the first scientific warning was published.)

All homes built before 1968 were therefore checked; 55,393 in all. The investigators reported that over 96 per cent of the communication pipes to that housing stock were made of lead, and the same was true for over 80 per cent of the supply pipes; in that community the average length of supply pipe was approximately 19 metres.

Lead was not being used to the same extent in the internal plumbing: approximately 27 per cent of the homes contained internal lead pipework.[142] Wherever lead pipes and tanks were found, the occupants were told to flush their taps to reduce their exposure to lead, and they were informed that grants for lead pipe replacement were available. The investigators were evidently disappointed that '...despite the publicity given to the survey during its operation in both local and national press and the fact that every house was left an information leaflet, public awareness as to the dangers of excess lead in drinking water remains disappointingly low.' They speculated that 'The reasons for public apathy are...due to the fact that lead in water is colourless, tasteless and odourless and is, therefore, not an emotive subject' and so observed that 'It is clear that official bodies – local authority or government – must take the lead.'[143]

The meaning of the final word in that quotation could be pronounced and interpreted in either of two ways. Do they mean that the government should set a bold policy or are they insisting that the old pipes should be removed? There was at least no ambiguity when they argued that the government should restore the rate at which grants for lead pipe replacement were paid to their former level of 90 per cent rather than the less generous figure to which it had by then been reduced. As recently as 1991, the Scottish Office estimated that 589,000 out of a total of 2.1 million dwellings, corresponding to 28 per cent of the housing stock, still had some lead plumbing. In the same year, the Drinking Water Inspectorate estimated that 3.1 per cent of water supplies in England and Wales did not meet the prevailing standard of 50 µg/L. Eight years earlier the Royal Commission on Environmental Pollution had pointed out that '...persons living in areas with plumbosolvent water and having lead plumbing or lead-lined water tanks may receive more than half of their uptake [of lead] from the water.'[144] Despite this, public investment in lead pipe replacement declined during the 1990s.

During the first half of the 1980s, all British water authorities complied with the requirements of European Directives, and surveyed their systems to check on levels of lead in tap water.[145] Those surveys indicated that the lead concentration in water supplies to approximately 5 million people in some 130 water supply zones (the vast majority of which were in four of a total of nine Water Authorities) remained consistently above 50 µg/L. The water supplied to those zones was therefore chemically treated with orthophosphates to reduce acidity, and thereby reduce contamination below 50 µg/L. Some of the highest levels of lead can be found in water supplied by South West Water, North West Water and Northumbria Water. As recently as 1989, there were some 20 water supply zones, serving more than a million people in total, who were then still receiving water which consistently contained more than 50 µg of lead per L.[146] In 1990, there were still 12 zones in which the water was being supplied with such high levels of contamination, yet there were no prosecutions, and no public warnings were issued.

The town of Blackburn in Lancashire provides an interesting but perhaps extreme example, because there, approximately

20,000 households receive water contaminated at significantly elevated concentrations, some as high as 50,000 or even 300,000 µg/L.[147] Approximately 57 per cent of around 2500 water samples collected in Blackburn in the ten years up to 1989 contained more than 50 µg/L of lead, and just over 30 per cent contained more than 500 µg/L.[148]

In a very small local study conducted in Blackburn in 1994, investigators found that during routine sampling about a quarter of all water samples had lead concentrations above 10 µg/L. When the water company received complaints specifically about drinking water being cloudy, samples were then collected and analysed. Of the 12 homes from which complaints were received, only two of the cloudy samples had lead concentrations below 10 µg/L, and only four were below 25 µg/L, while the highest contained 88,200 µg/L.[149] The investigators acknowledged that 'There are...no data available on the frequency of high dissolved lead content at any individual households.'[150] That is true not only for Blackburn, but also more generally. We know that all too often the concentration of lead in water is too high, but we do not know how frequently the problem is occurring.

In 1989, Blackburn Borough Council provided approximately 300 grants to help householders meet the cost of lead pipe replacement, but in February 1990 the Department of the Environment terminated that grant scheme. North West Water has reported that in 1995 they replaced 30,105 lead service pipes in response to requests from their customers, and just 10,754 on their own initiative, which makes a total of just over 40,000.[151] They state, however, that on average they usually replace more than 50,000 lead service pipes each year, but why fewer were replaced in 1995 remains unclear. The company was unwilling to forecast a date by which all lead pipes would be removed, saying that '...too many uncertainties surround this issue.'[152]

Because of the problems in Blackburn, and the slow rate at which leaded domestic plumbing was being replaced, the local borough council persuaded North West Water to establish a scheme under which the water company would provide interest-free loans to local residents if they had their internal lead water pipes replaced at the same time as lead was being removed from the external service. The scheme was introduced in the early summer of 1996, but by August only a few hundred householders had taken advantage of it. At current rates, therefore, the Environmental Health Department of Blackburn Borough Council estimated that it would take over 20 years before all the leaded pipes were replaced.[153] This new scheme replaced an older one under which home owners had to meet the full cost of replacing the internal lead pipes before replacement work would start. Under that arrangement, the eradication of lead piping would have taken considerably longer.

The circumstances in Blackburn are unusual but the chemical characteristics of the water are sometimes such that lead particles dissolve into, or are dislodged into, the drinking water irrespective of its pH. This is continuing to happen even though the water supply has been treated with lime for many years to reduce plumbosolvency.[154] Treatment to reduce acidity is therefore not by itself sufficient to solve the problem of lead in drinking water. Indeed Quinn and Sherlock estimated in 1990 that when all the then planned water treatments had been implemented, there would still be at least 4 million dwellings in England alone in which average water lead concentrations would regularly exceed 10 µg/L.[155]

It is therefore not entirely surprising that scientists at the WRC concluded in 1992 that if drinking water in Britain is to meet the World Health Organisation's new maximum lead contamination limit of 10 µg/L the only practical solution would be the replacement of all lead pipes.[156] During a set of hearings by the House of Lords Select Committee on the European Communities it emerged that this crucial fact is appreciated both by the water companies and by the Select Committee, but the government has yet to explicitly acknowledge it.[157]

In 1992 the WRC provided the Department of the Environment with a report on the costs of lead-pipe replacement in which the authors estimated that there were some 8.9 million properties in England and Wales supplied with lead pipes owned by the water companies. They also reported that they had estimated that each year approximately 1.5 per cent of households replace their lead pipes.[158] The House of Lords Select Committee therefore reminded the government that the Royal Commission had advised as long ago as 1983 that financial considerations should not to be allowed to hamper the lead pipe replacement programme.[159] To suggest to the Conservative government that financial considerations should not hamper their actions, when it came to protecting public health, was to speak a language which ministers insisted on pretending that they could not or did not understand.

Quinn and Sherlock went slightly further, in 1990, when they observed that to meet a water standard of 10 µg/L, not merely would it be necessary to replace all lead pipes, but 'In addition, capillary lead-soldered joints for copper pipes should no longer be used in either institutional buildings or domestic potable water plumbing systems.'[160]

It remains to be seen what steps will be taken not just to

reduce plumbosolvency of drinking water, but to replace leaded pipes, and to ensure that leaded solder is not used in drinking water systems.

Lead in paint, dust and soil in the UK

As early as 1904 Dr Lockart Gibson, a physician in Queensland, Australia, concluded that lead paint in the home was responsible for lead poisoning in children. He understood even then that the problems occurred when the paint deteriorated and fragmented, and especially when babies and young children played with toys and other objects which went into their mouths, carrying the hazardous dust.[161]

When compared to the US, far less is known about the extent to which lead paint is contaminating house dust in the UK, and this is directly connected to the fact that public policy has been singularly negligent in this regard, even when compared with the generally negligent conduct of British authorities, both in relation to general issues of environment and health and in respect of other lead-related issues.[162] The description given above, in the section on lead in paint in the US, explains just how detailed a picture US authorities have assembled and published. This is, no doubt, also connected with the fact that American authorities have developed a comprehensive policy and strategy for the identification and abatement of lead-contaminated housing stock.[163] US regulatory authorities have not only specified maximum levels of lead above which lead abatement is considered necessary, but they have done so both for percussive surfaces, by which they mean those which receive regular impacts such as door and window frames, and for non-percussive surfaces, such as walls and ceilings. They have also funded research into improving and evaluating the costs and performance of methods for removing or immobilizing old leaded paint.[164] But then US government departments also fund programmes to identify, remove or encapsulate old leaded paint in residential households and to train people to perform and supervise those activities.[165]

In 1980 the Lawther Committee took some comfort from the fact that just 3 per cent of the paint then being sold in the UK had a high lead content. That represented, however, the use of some 1500 tonnes of lead annually.[166] The government, in the form of the Department of Health, has consistently denied that British children are at risk from this source of lead exposure, and survey work is not taking place.[167] Those denials have rather drowned out the warnings that have been given, including one from the Royal Commission on Environmental Pollution in 1983.[168] The Commission then recommended that 'The government, in collaboration with the paint industry, should establish the quantities of lead-based paints currently sold in the UK.'[169] The government responded with an announcement that the Paintmakers Association of Great Britain had agreed to carry out a survey of its members to establish the quantities of paint made in the UK. They proposed to collect statistics under four headings: paints containing no deliberately added lead, those with lead levels below 0.5%, those with lead levels between 0.5 per cent and 1 per cent, and those above 1 per cent. The Department of Trade and Industry meanwhile was supposed to compile estimates from those paint manufacturers who were not members of that association.[170] There is no evidence that investigation was ever conducted, and if it has, the results have never been published.[171] The government pointed out that the Paint Research Association had commissioned research to find alternatives for most uses of lead in paint, indicating in effect that they were not willing to set any standard until after they were sure that it had already been met.[172] What they have never done is to find out just how many homes have contaminated paint and dust, or if they have gathered that information, they have never published it. The British government has, moreover, never initiated any programme of lead paint abatement, nor has it taken any steps to encourage paint abatement, and no standards for paint abatement have ever been set in the UK.

The most comprehensive study on lead levels in dust and soil in the UK was commissioned by the Department of the Environment and conducted by a team based at Imperial College in London.[173] The results were published in detail in 1990, but the samples were collected between November 1981 and June 1982. Dust and soil samples were carefully collected in 53 locations in England, Scotland and Wales, including seven London boroughs, 'hot-spots' such as old Derbyshire mining villages, and some rural locations. Samples were taken at 100 homes in each of those 53 locations and some 4600 samples of house dust and 4100 samples of soil were analysed. Samples were also collected of dust from school playgrounds, and soil from gardens, including plots used to grow vegetables. The report of the study gives a more comprehensive account of levels of lead contamination than has ever been provided before or since.

The main results for levels of lead in dust and soil in 53 cities, towns and boroughs, were represented by the investigators in the graphical form shown as Figure 4.9. For each location, a horizontal line is provided; the small dark circle in the middle of those lines represents the geometric mean of the

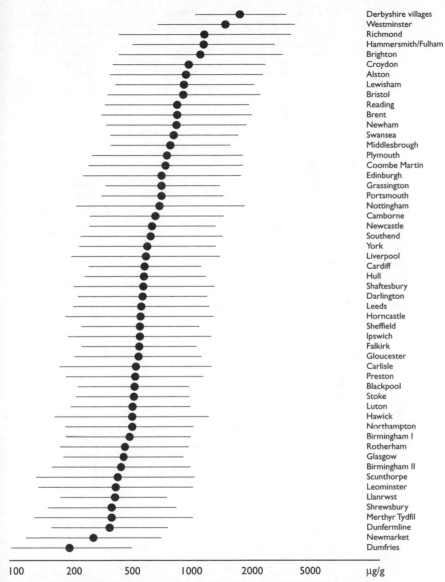

Derbyshire villages
Westminster
Richmond
Hammersmith/Fulham
Brighton
Croydon
Alston
Lewisham
Bristol
Reading
Brent
Newham
Swansea
Middlesbrough
Plymouth
Coombe Martin
Edinburgh
Grassington
Portsmouth
Nottingham
Camborne
Newcastle
Southend
York
Liverpool
Cardiff
Hull
Shaftesbury
Darlington
Leeds
Horncastle
Sheffield
Ipswich
Falkirk
Gloucester
Carlisle
Preston
Blackpool
Stoke
Luton
Hawick
Northampton
Birmingham I
Rotherham
Glasgow
Birmingham II
Scunthorpe
Leominster
Llanrwst
Shrewsbury
Merthyr Tydfil
Dunfermline
Newmarket
Dumfries

100   200   500   1000   2000   5000   µg/g

Each bar shows the geometric mean ±1SD (ie, 68 per cent range) for 100 houses.
Source: I Thornton et al (1990) 'Lead exposure in young children from dust and soil in the United Kingdom' *Environmental Health Perspectives*, vol 89, pp56–57.

Figure 4.9a: *Concentrations of Lead in House Dust*

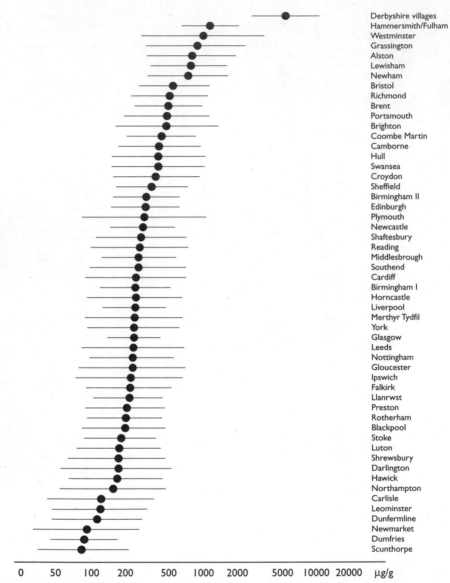

Each bar shows the geometric mean ±1SD (ie, 68 per cent range) for 100 houses.
Source: I Thornton et al (1990) 'Lead exposure in young children from dust and soil in the United Kingdom' *Environmental Health Perspectives*, vol 89, pp56–57.

Figure 4.9b: *Concentrations of Lead in Garden Soil*

sample, while the length of the side arms indicates the magnitude of the standard deviation of those means for those samples. The full distance of those lines, from left to right therefore (in accordance with the statisticians' definition of a 'standard' deviation) represents 68 per cent of the range, but not the entire range. The investigators also provided a table that serves to highlight their main conclusions, and most of their figures are reproduced in Table 4.6.

Table 4.6 *Lead concentrations in survey locations derived according to sample type*

Sample type	All study locations except hotspots	London boroughs	Derbyshire mining villages	Remaining geochemical hotspots
House dust				
Number of samples	4638	683	100	492
Geometric mean (µg/g)	561	1010	1870	631
Range (µg/g)	5–36,900	5–36,900	606–7020	74–40,300
Playground dust				
Number of samples	220	34	5	18
Geometric mean (µg/g)	289	430	4390	400
Range (µg/g)	11–6,860	93–6860	1190–13,400	53–21,700
Garden soil				
Number of samples	4126	578	89	433
Geometric mean (µg/g)	266	654	5610	493
Range (µg/g)	13–14,100	60–13,700	1,180–22,100	49–8340

Source: I Thornton et al (1990) 'Lead Exposure in Young Children from Dust and Soil in the UK' *Environmental Health Perspectives* Vol 89, p58, Table 1

Thornton and his colleagues found, for example, that the geometric mean level of lead in house dust, in areas other than industrial hotspots, was 561 µg/g, and they also reported that overall 10 per cent of house dust samples exceeded 2000 µg/g as did 5 per cent of soil samples.[174] That relatively high bench-mark was, moreover, exceeded in 18 per cent of London homes and in 44 per cent of those sampled in Derbyshire. The authors did not pull their punches. They made it clear that 'This national survey clearly confirmed that lead contamination is widespread in both soils and dusts in Britain and establishes an urgent need to assess the relative importance of lead-rich dust and soil as a source of lead exposure to young children.'[175] For British civil servants, even when in the company of leading academics, to publish a sentence as blunt as that is truly remarkable, even in an academic journal.

Despite those findings, the Department of the Environment's support for research into the levels of lead contamination in soils and household dust in the UK has been curtailed.[176] Moreover, although the Department of the Environment had a Steering Committee on Environmental Lead Monitoring, not a single meeting of that Committee has taken place since 1988.[177]

A second related research project was conducted by Davies et al in Birmingham between 1984 and 1985, and it was jointly funded by the DoE and MAFF. This study investigated just one city, but it did so in considerable detail. The Birmingham study provided an important advance on the previous project, because in this case they did not just estimate lead concentrations in the environment, but they also measured blood lead levels in young children, in this case two-year-olds. The investigators estimated that the geometric mean blood lead level was 11.7 µg/dL, and that 5 per cent of the children had PbB levels above 24 µg/dL.[178]

The authors remarked that their Birmingham study '...has for the first time shown a significant relationship between levels of environmental lead within the home ...and blood lead in the 2-year-old child.'[179] It may, strictly speaking, have been the first time that this relationship had been demonstrated from UK data, but it had already been demonstrated elsewhere, especially in the US, and in the work of Steenhout in Belgium.[180]

The data from the Birmingham study provide a unique opportunity, as far as published data are concerned, to compare the results obtained in blood lead measurements, with those which would have been predicted by the US EPA's Uptake and Biokinetic model. When the data for lead levels in key environ-mental media are introduced into the UBK model, it predicts a geometric mean PbB figure of 9.7 µg/dL; ie a figure which is just less than 20 per cent below that which was actually measured. In the circumstances, therefore, it would seem prudent, until more data are published, to interpret the predic-tions of the UBK model as, generally speaking, underestimating blood lead levels by approximately one-fifth of their true value.

Towards the end of their report Thornton and his colleagues also remarked that 'If the action limit value for blood lead of 25 µg/dL were to be lowered to say 15 µg/dL this...would have considerable implications, and many British households would require either some form of remedial action or advice on cleaning procedures.'[181] This is as close as UK officialdom has ever come to acknowledging one of the central claims of this book. They are right. Many British households do require 'some form of remedial action' and rather more 'advice on cleaning procedures'.

At a local level, environmental health officers in the UK do monitor some paint work for lead. The data available are not sufficiently detailed to give any idea of the adequacy of the monitoring or of the scale of lead contaminated dust and paint-work. For example, between 1987 and 1990 only about 30 local authorities out of a total of 403 in England and Wales, investigated the lead concentration of old paint each year. Levels of lead in dust were also investigated in 49 local authorities in 1989, 80 local authorities in 1988 and 58 local authorities in 1987. In 1991 the Chartered Institute of Environmental Health wrote to all 403 local authorities in England and Wales; 300 of those replied, and just 22 had done some monitoring work on heavy metals in paint.[182] As one Environmental Health Officer remarked, 'We are simply not resourced to undertake local monitoring if there is not a particular medical problem brought to our attention.'[183]

In 1992 the *British Medical Journal* reported on a study conducted by two French scientists who were working in Paris for Médecine Sans Frontières.[184] After two children died in Paris from lead poisoning, they initiated a national survey to look for further evidence of lead poisoning. After surveying just three districts in Paris they reported that they had found 1500 cases of lead poisoning. This surprised them as lead-based paints had twice been banned in France, firstly in 1913 and again in 1948. Following the study, treatment that was administered to reduce the lead contamination was very effective in reducing levels of lead in both house dust and in the children's blood.[185] The efforts of Médecine Sans Frontières show that work to locate leaded paint and to diminish the residue is no less worthwhile in Europe than it is in the US, and if they can do it in Paris why can't it be done in Plymouth or Portsmouth?

Lead in air in the UK

The main source of airborne lead has for many years been the exhausts of vehicles fuelled by petrol containing lead compounds. The levels of lead used in fuel started to decline in the early 1970s, and since 1976 the total quantities of lead released on British roads has been diminishing. In 1976 approximately 8400 tonnes of lead were emitted by British vehicles; by 1990 the annual figure had fallen to close to 2000 tonnes, while by 1994 it was approximately 1300 tonnes.[186] It was not until 1981, however, that levels of airborne lead started to decline. Unleaded motor fuel was introduced in Britain in 1986.

Official interest in this topic can be traced back to the early 1970s when the Department of Health's Chief Medical Officer advised the British government that what was then being experienced, namely a continual and rapid increase in lead emissions from vehicle exhausts, could not be sustained.[187] The problem was that the number of vehicles on the road was rising rapidly and so was the average number of journeys for which those cars were being used. The British government responded to those considerations in 1972 by committing itself to ensuring that the concentrations of lead in air should, in the long run, not exceed those which prevailed in 1971.[188] The level at which lead could be added to petrol was reduced in five steps going from 0.84 g/l in 1973 to 0.4 g/l in 1981, and since 1986 the maximum figure has been 0.15 g/l. The pattern of those reductions is illustrated in Figure 4.10.

Despite those reductions in the amount of lead added to each gallon, the overall quantity of lead emitted from the British vehicle fleet did not decline to 1971 levels until 1981.[189] By 1984 all Member States of the EEC had agreed to ensure that unleaded petrol would be made available throughout the Community by 1989 and they also undertook to 'promote its widest possible use'.[190]

In 1983, the Royal Commission on Environmental Pollution estimated that lead from vehicle exhausts was responsible for approximately 20 per cent of the body lead load in Britain, and indicated that it was their view that it should be diminished.[191] Several surveys of levels of lead in air in Britain have provided evidence to show that the levels of airborne lead have been declining since the early 1980s, especially since the sale of lead-free petrol began.[192] For example, a Department of the Environment report indicated that between 1984 and 1987 reductions at various urban and rural sites in Wales ranged between 52 and 61 per cent.[193]

The illustration provided in Figure 4.11 shows that lead emissions from British motor vehicles have fallen over the last 20 years, and the sharpest declines followed the reductions in the levels of lead in petrol in 1981 and 1985. The period from 1976 to 1990 was, nonetheless, one which saw an increase in

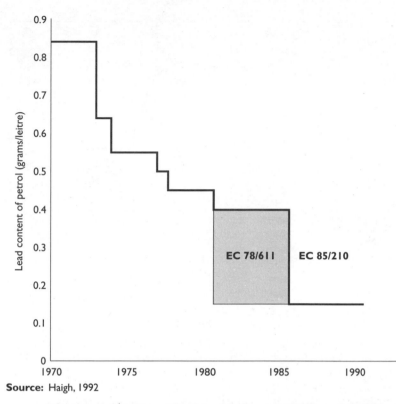

Source: Haigh, 1992

Figure 4.10: *Permitted Lead Content of Petrol in UK*

petrol consumption in the UK of approximately 44 per cent. Lead-free petrol first became available in the UK in June 1986. The UK was in this regard a very long way behind the US, since by 1980 over 90 per cent of gasoline sold in the US was already unleaded. The adoption of unleaded petrol was, however, initially sluggish.[194] Even as late as October 1988, only 3.5 per cent of fuel outlets sold unleaded petrol and unleaded petrol sales amounted to only 1 per cent of the total UK petrol sales.[195] The government's 1987 budget had introduced a small tax differential in favour of unleaded petrol but it was not sufficient to persuade most motorists to switch to unleaded fuel. Sales of unleaded petrol rose rapidly after 1988 following an advertising campaign and several extensions to that tax differential.[196] Unleaded fuel did not account for 20 per cent of sales until the summer of 1989, and it did not reach 40 per cent until the spring of 1991.[197] The 50 per cent mark was not reached until the start of 1993, and by the autumn of 1996, unleaded fuel accounted for just over 68 per cent of all the petrol sold in the UK.[198]

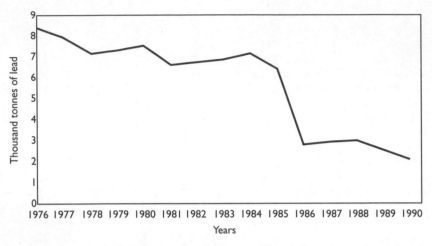

Figure 4.11: *Estimated emissions of lead from petrol-engined road vehicles*

Lead in food in the UK

The earliest indication of the British government's interest in food-borne lead can be found in a study supported by the Ministry of Health which was published in 1938.[199] On that occasion Monier-Williams combined the data on food consumption which the Ministry had gathered in 1921 with his estimates of the levels of lead in various types of foods, and concluded that average adult daily intakes of lead were between 0.2 and 0.25 mg/day.[200]

In 1950 the Ministry of Agriculture, Fisheries and Food (MAFF) repeated the exercise. On that occasion, they used the results of the 1949 National Food Survey, and concluded that average adult intakes had by then risen to between 1 and 1.5 mg/day. That information was not published at the time; indeed it was not released until 1982, and even then it only emerged in the very last two lines of a relatively obscure official report.[201]

In 1982 MAFF implicitly acknowledged that for much of this century, levels of lead in British diets have been significantly higher than would now be considered prudent.[202] In 1983 the Department of the Environment stated that most of the lead which is taken in by humans originated in their food, but those are words which MAFF has never brought itself to use.[203]

In 1944 the British government and the canning industry thought that it was extremely difficult to manufacture canned corned beef containing less than 2.5 mg of lead per kg. Indeed in the 1940s some samples of corned beef were found to contain levels of lead as high as 30 mg/kg, while almost 30 per cent of the cans sampled contain more than 5 mg of lead per kg.[204] By the early 1950s the position had actually deteriorated, and some samples were found to contain as much as 45 mg of lead per kg, and one-third having more than 5 mg of lead per kg.[205] Canned fish (at least until the late 1950s) was also frequently highly contaminated with lead, especially sardines, pilchards and shellfish.[206] Before the Second World War it was even common practice to grill sardines on lead-coated trays.[207] Until the 1950s, moreover, it was customary for tea leaves to be packed in lead-lined chests, and in 1949 it was not unusual for tea to contain as much as 28 mg of lead per kg.[208]

In areas of Britain where lead has been mined or smelted, such as North Derbyshire, very high levels of lead have been found to contaminate the soils, and consequently some food products grown on those soils.[209] Fodder crops grown in that area were found to be highly contaminated with lead, which in turn made its way into the milk of cattle that consumed the fodder.[210] Consequently, steps were taken in the late 1970s and early 1980s to reduce that localized problem.[211]

Traditional canning technology used soldered cans in which the solder consisted of 98 per cent lead and 2 per cent tin, but by the early 1980s two new types of can-making technology were being introduced with either two-piece cans or welded cans. By the mid-1980s the major canning companies in both the US and the UK were using these new technologies, but the same could not be said of the smaller companies, nor of overseas suppliers. As recently as 1986 a study conducted in New Zealand found that while canned foods accounted for an average of 8.4 per cent of the diet, they provided more than 34 per cent of dietary lead. Similarly, in the UK in 1982 canned vegetables constituted only 2.1 per cent of the diet but 10 per cent of dietary lead intake.[212]

In 1982 MAFF published the results of its food surveillance programme on samples collected between 1977 and 1981.[213] That study might be thought to have been an exceptionally comprehensive one, since over 3000 food samples were tested from some 134 diets.[214] A more careful scrutiny of the report reveals, however, that those figures represent the sum of their efforts over seven years, and that the highest number of diets examined in any one year had been just 25 in 1977, and in the most recent year, 1981, just eight diets had been studied. Each diet was, moreover, studied for just a single day. When viewed in that light, the project then looks rather less comprehensive.[215] MAFF reported that in those modest, and possibly unrepresentative, sample diets, the average lead intakes had declined from 0.12 mg/day in 1975 to 0.08 mg/day in 1981. These results were obtained by scrutinizing a small sample of adult diets and an obvious omission was the absence of specific intake estimates for young children.

When the Lawther Committee reported in 1980 to the British government they did not mince their words when they stated that 'The evidence...on human exposure to lead indicates that, for the population as a whole, the diet is the most important route of exposure.'[216] When the Lawther Committee discussed levels of lead in the British diet they reported their estimates in terms of µg/day, but when the Ministry of Agriculture, Fisheries and Food dealt with the same topic it used units one thousand times as large, namely mg/day; the latter units have the virtue that when reporting the same measurement, the estimates are prefaced by seemingly lower numbers.

In their 1982 report, MAFF acknowledged that cans of acidic foods, such as blackcurrants, fruit juices and tomatoes, contained particularly high concentrations of lead.[217] They calculated that the average diet contained 6.2 per cent by weight of canned food, and just over 0.5 mg/week of dietary lead, but if the proportion rose to 39 per cent by weight of canned food then lead intake would rise to 1.6 mg/week, which corresponds to what was then three times the population average.[218]

MAFF published its next review of levels of lead in food in 1989, when results obtained between 1982 and 1987 were published.[219] MAFF estimated that by then average adult lead intakes had declined to between 0.02 and 0.07 mg/day, although that excluded intakes from drinking water, and once again no corresponding figures for young children were provided.[220] By 1987, levels of lead in most dietary components were reasonably low, but higher levels continued to be found in meat offals, mussels and in canned foods, although even in those products the levels found were lower than those reported in 1982.[221] In 1991 MAFF indicated that its Total Diet Study suggested that by then average lead intakes had declined to 0.015–0.028 mg/day.[222] Since, however, the Total Diet Study did not include food or beverages consumed outside the home, or tap water, it must be assumed that those figures represent underestimates rather than overestimates.

Although it remains difficult to estimate the amounts of lead being ingested by young children, or even to know how representative were the adults' diets studied by MAFF, the reductions in food-borne lead in the UK over the last 30 years have been quite striking. It is ironic that the one area in which the British authorities have chosen to take regulatory action is the one in which the American authorities have relied on persuasion rather than compulsion. It is also ironic that the regulatory action taken by the Ministry of Agriculture, Fisheries and Food in relation to food-borne lead has not merely been more robust than that adopted by the FDA, it has also been markedly more rigorous than MAFF's policy on almost any other aspect of food safety. When the London Food Commission published its comprehensive review of food adulteration in Britain in 1988, it made no reference whatsoever to lead contamination.[223] The main reason why lead was not discussed in that text was because lead was one of the few areas of food safety policy with which the British authorities *had* been dealing diligently. The authors therefore concentrated on the numerous other aspects of food safety policy which needed to be strengthened. By comparison with drinking water and lead-based paint, food is not a major source of lead exposure in the UK.

Conclusion

The evidence reviewed in this chapter shows that the US provides far more detailed information than the UK, but that in both countries average blood lead levels have been falling since the early 1980s. The rate of decline in the US has been more rapid than that from which the British population has benefited. In 1994, the evidence from the NHANES III study suggested that just under 10 per cent of American children aged five or less have PbB levels elevated to about 10 µg/dL. Far less precision is possible in relation to the UK, but in the mid-1980s, something in the order of a quarter of all children in the UK probably had PbB levels above 10 µg/dL. By the mid-1990s, that proportion had declined, but it remains likely that somewhere between 10 per cent and 15 per cent of all young children in the UK have PbB levels elevated above that value.

Public health policy, however, needs to pay attention not just to average PbB levels or even to the proportion exceeding any particular bench mark, but developments in toxicology also need to been taken into account. Toxicologists have provided evidence of adverse effects at ever lower PbB levels, and no evidence of a threshold has been found. This means that the policy-makers need to aim at an ever more demanding target, and so it would be premature for the British government, in particular, to think that the problem of lead pollution has been solved.

Average blood lead levels have declined, but a significant proportion of young children remain vulnerable to hazardous exposures. Because old paint and drinking water are the most common sources of lead pollution for the vast majority of those over-exposed, the groups in most need of help are most likely to be found in the older housing stock which contain lead-based paint and are serviced by lead water pipes. These children need and deserve help which they are not yet receiving.

5

Government Strategies for Controlling Lead

An important defining characteristic of US government policy on lead and health is that it has been spelled out in detail in a public document which has been widely distributed. In 1991, and again in a slightly revised form in 1993, the EPA published a document defining the federal government's strategy for reducing human exposure to lead.[1] The policy is, moreover, remarkably clear, straightforward and sensible. The main barrier that stands between articulating the strategy and solving the problem is that Congress needs to provide the resources for the strategy to be fully implemented.

US government strategy

The EPA strategy document asserts that the Agency's primary goal is to reduce lead exposures to the fullest extent practicable, and especially to reduce exposure for young children, while taking into account the associated costs and benefits.[2] In other words, the government is not just proposing to cut exposures sufficiently to reduce the blood lead level of all children down, for example, to 10 µg/dL, and for adults down below 25 µg/dL. The government proposes to continue to diminish exposures, but it does not think it realistic to eradicate lead entirely from our environment. The Agency understands, moreover, that the costs of continuing to diminish lead exposure can escalate unrewardingly. It would be far more expensive to cut childhood blood lead levels from 5 µg/dL to zero, than it would be to reduce them from 10 µg/dL to 5 µg/dL.

The EPA's strategy is not just realistic, it is also practical. The main elements of the strategy are to:

- develop methods to identify geographic 'hot spots';
- implement a lead pollution prevention programme;
- strengthen existing environmental standards;
- develop and transfer cost-effective abatement technology, especially for dealing with leaded paint;
- encourage the availability of environmentally sound recycling of leaded wastes;
- develop and implement a public information programme; and
- aggressively enforce environmental standards.

If all those goals are achieved, the adverse effects of lead on public health in America will be substantially diminished. Remarkably enough, although recent years have seen some fierce debates in Congress about many of the EPA's policies and actions, the EPA's policy in respect of lead has not been significantly challenged in, or undermined by, Congress.

One of the reasons why that is so is that the toxicological evidence is so robust that it has, if anything, strengthened rather than weakened in recent years. Another reason is that most current uses of lead, especially in batteries, are not contributing significantly to environmental exposures, except for the communities located close to smelters and battery plants. Those firms who are currently active in producing and using lead are not therefore the main targets of regulatory action, the focus is almost exclusively on clearing up the residues of old uses.

The three major sources of excessive lead exposure in the US today are old lead-based paint, urban soil and dust (contaminated old paint and vehicle exhausts), and drinking water contaminated by old lead pipes and lead-containing plumbing fittings, and the EPA is, understandably, focusing its energies on those main routes of exposure.

In practice, however, as Goldman and Carra pointed out in the *New England Journal of Medicine* in July 1994, following the publication of the results of the NHANES III study, the available evidence indicates that 'US strategy must begin to focus more than ever on poor, nonwhite, and inner-city children. We must intensify our efforts to screen these children, identify the sources of lead exposure, and eliminate or reduce these exposures.'[3] That task will be harder to complete than to describe, however, because in the US, as Goldman and Carra acknowledge, 'Many of these children do not have access to routine medical care....'[4] This is one crucial respect in which children in the UK are better placed than those in the US. There are many shortcomings with Britain's National Health Service, but one simple virtue is that the system is able to provide all children with publicly funded professional medical care – free at the point of delivery. If the UK government chose, therefore, to test blood lead levels in all young children, it would be a far more straightforward task in Britain than in the US.

A challenge which confronts the lead industry in both the UK and the US, and elsewhere, is whether it can safely develop new products containing lead, and new uses for lead. New uses continue to be introduced, but there is no requirement in either country for any regulatory scrutiny of lead-containing products prior to marketing. The EPA's Office of Toxic Substances is, however, considering issuing a rule requiring advanced notice for anyone intending to manufacture or process lead for new uses, '...in order to ensure that those uses do not pose unreasonable risks.'[5] Whether or not prior approval becomes a formal requirement, it would be commercially imprudent for a company to try to market any product which might attract public criticism.

UK government strategy

As far as the UK government is concerned, no comparable strategy has ever been articulated or published. The British authorities have far less in the way of an explicit strategy. It would be a mistake, however, to conclude that no strategic thought has ever been devoted to this topic within British government departments. A few fragmentary strategic remarks have been uttered by ministers and mentioned in documents, though UK policy has never been as comprehensive or as transparent as that of the US. The evolution of British policy can be pieced together because some actions rather than none have occurred, and some targets rather than none have been set. The evidence suggests, however, that one implicit rule of UK policy has always been, in effect, never to set a target until it has already been met, and therefore never to indicate to the general public that there is a widespread or serious hazard, or that urgent action is required.

For many years British policy was limited to trying to restrict lead exposures for those working in the lead industry. In 1920, for example, parliament passed *The Lead Processes Act*, which made it unlawful to employ women and 'young persons' for a wide range of industrial processes such as furnaces, refineries, foundries and the manufacture of solder. Under that legislation women and childen were also banned from cleaning the workrooms where any of these processes were conducted. In 1926 the *Lead Poisoning Act* supplemented those provisions by setting restrictions on the conditions under which people could be employed 'in connection with the painting of buildings'. Remarkably enough, that Act prohibited '...dry rubbing down and scraping'.[6] What is remarkable is that, 70 years on, dry sanding and scraping remain normal practice throughout the UK, both amongst professional painters and decorators and amateurs.

The 1926 Act also required '...the use of protective clothing...' and '...the distribution to persons so employed of instructions with regard to hygienic precautions to be taken.'[7] In the years since then, the levels of lead in paint have been sharply reduced, but the reductions in the levels at which known adverse effects occur have been reducing almost as rapidly, and therefore the precautions which were required in the 1920s are no less relevant today – the main problem is that the people who need to know are not being told.

There have been occasions when there has been pressure on the British government to address problems of lead exposure, but the official responses have never been as prompt or as comprehensive as the British public might have wished. The reviews in Chapters 3 and 4 showed that in the late 1970s there was evidence of high levels of exposure in Britain and strengthening evidence of the toxicity of lead, while the Department of Health and Social Security responded not by taking any regulatory initiatives, but with that classical delaying tactic: a committee of enquiry was established.

The working party chaired by Professor Patrick Lawther was set up in 1978 and reported in March 1980. Its terms of reference were 'To review the overall effects on health of environmental lead from all sources and, in particular, its effects on the health and development of children and to assess the contribution lead in petrol makes to the body burden.'[8] When the Lawther report was eventually published on 28 March 1980 the working party concluded that 'We have not been able to come to clear conclusions concerning the effects of small amounts of lead on the intelligence, behaviour and performance of children.'[9] That statement was highly controversial, and was subsequently repudiated by some members of the committee.[10] The Lawther committee were particularly criticized by Bryce-Smith and Stephens for failing to take account of evidence from biochemistry or from animal studies.[11] Indeed they characterized the Lawther Report as '...too deeply flawed and fragmentary to merit further serious consideration.'[12]

Whatever the merits or shortcomings of the Lawther committee's report, the important fact was that it did play an influential role in British policy-making, if only in a negative sense. In practice the committee allowed its judgement, that there was no conclusive proof that low level lead exposure was causing adverse effects, to be interpreted by the government as if it had provided conclusive proof that no such adverse effects were occurring. The report of the Lawther committee provided the British government with an excuse for doing as little as possible, as slowly as possible.

The Lawther committee thought that, within its own terms, it was taking a cautious approach, because it recommended that if a child was found to have a PbB level above 35 µg/dL then steps should be taken to ascertain the source of exposure, and to reduce it.[13] In practice, however, that was not a particularly cautious judgement, firstly because the evidence which was then available would have justified specifying a PbB figure significantly below 35 µg/dL, and secondly because that benchmark allowed for no margin of safety whatsoever. Furthermore, the committee did not recommend the establishment of a screening programme to identify those children with elevated PbB levels. The committee did, however recommended that 'There should be a programme for the detection of lead in paint coatings accessible to children in areas where a high incidence of old lead paint surfaces may be suspected, such as old inner city residential areas.'[14] That recommendation was sensible, but has still not been implemented.

The Lawther committee did argue that the levels of lead in drinking water should be reduced, but again the wording was ambiguous. The report states that 'The reduction should generally be to the maximum extent possible by reasonable means.'[15] But that phrase is crucially ambiguous as between adding chemicals to reduce plumbosolvency on the one hand and the complete removal of all lead pipes, tanks and fittings on the other. The committee did, however add that

> ...it should generally be recommended that people in affected households should avoid drawing the first run-off for drinking or food preparation after cold water has stood in the pipes for a few hours or more. Water from hot taps should not be used for these purposes, especially not for preparing infants' feeds. Special attention should be given to affected households where there are babies or expectant mothers and it may even be necessary to make changes to plumbing in an individual dwelling or to supply safe bottled water.[16]

That advice was prudent then and remains sensible, but the British government has not followed it. On the contrary, as will be explained below, the government has cautioned water companies and local authorities against passing that advice to the public without first considering the commercial interests of the water industry.

The Lawther committee also remarked that 'We have been told that early replacement of lead pipes on a large scale is not a practical solution due to costs and the extent of the requirement for trained labour and suitable material.'[17] The committee did not say who told them so, but the government must be the most likely suspect. We are entitled to ask, however, why the committee tamely accepted this attempt to invoke economic and political considerations to colour their scientific judgement instead of contesting it.

The response of the British government has been just about as ambiguous as the Lawther report itself. In 1982 the Department of the Environment (in collaboration with the Welsh Office) issued a circular to local authorities.[18] The circular started with an uncharacteristic acknowledgement that 'It is important that there should be an adequate margin of safety in relation to the amounts which people take in. The Government's view, in the light of the [Lawther] Report is that the safety margin may at present be too small...'[19] The Department then took the extraordinary step of going beyond the Lawther committee's recommendation by stating that '...where a person – *particularly a child* – is confirmed to have a blood-lead level over 25 µg/dL, his or her environment should be investigated for sources of lead and steps taken to reduce exposure.'[20] (emphasis in the original). That implies that the government then thought that a safety margin of 10 µg/dL should be provided between levels of actual exposure and the level at which there was convincing evidence of a hazard. If that same judgement were now to be applied it would entail that exposure should be totally eradicated. Be that as it may, the fact remains that a target was then in effect set, but very few actions have subsequently been taken either to ensure that it is met, or to revise it in the light of subsequent scientific developments.

The Department asserted that 'Government policy...is (i) to tackle local environmental "hot spots" where exposure to one source or several is likely to give rise to blood-lead levels over 25 µg/dL, and (ii) to seek to reduce exposure generally as far as is reasonably practical.'[21] That is a commitment of which the government has almost entirely lost sight. It is not that the government has subsequently done nothing, but rather that its actions have been slight, discreet and half-hearted. Action has beeen taken, but almost all developments in UK policy to protect the public from lead exposures since 1982 have been initiated by developments which took place outside the UK, primarily from the institutions of the European Community such as the Commission and the Council of Ministers, but also from competition between the British motor industry and the major overseas suppliers, and to a lesser extent from the World Health Organisation.

More importantly, since the US EPA published its magisterial four-volume study on *Air Quality Criteria for Lead* in 1986, and the results of the five major prospective studies, which were discussed in Chapter 3, were published, along with the report of the World Health Organisation's IPCS Working Party on Lead, the silence on this topic from the British government has been deafening, and its lack of overt action has become increasingly culpable.

The UK government has no strategy for dealing with old leaded paint, and having sold the water companies into the private sector in 1989, it has effectively washed its hands of responsibility for lead in drinking water. The blood lead action level remains at 25 µg/dL, even for young children, and no official comments whatsoever have beeen made on the numerous advances in lead toxicology which have taken place since the early 1980s.

The Department of the Environment has hardly said anything about lead since the start of this decade. In 1990 the Department published a document entitled *This Common Inheritance*, which was a commentary on the government's White Paper on the environment. Its references to lead were contained in one sentence, namely, 'Levels of lead in air have been substantially reduced by effective action to control pollution from industry, and by reductions in the use of lead in petrol, encouraged by tax incentives for unleaded fuel.'[22] Her Majesty's Inspectorate of Pollution may have taken action on industrial emissions, and the growing use of unleaded motor fuel has contributed to lower levels of lead in the air, but the absence of any reference to old leaded paint or to lead water pipes is hard to justify.

In 1995 the Department indicated that it was minded to invest in research on the potential human and environmental exposures which could arise from lead's current use as an additive in paints, primers, plastics, ceramics, glass and alloys. What was so curious about that initiative was that it focused exclusively on the most minor sources of exposure while entirely neglecting the major sources. The Department said that it wanted to update its knowledge, but the topics it chose to study could not be explained by reference to any toxicological or public

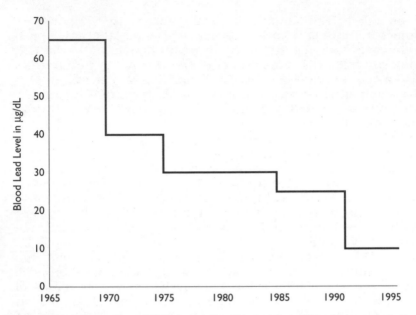

Source: Centre for Disease Control (1991) *Preventing Lead Poisoning in Young Children: A Statement by the Centre for Disease Control,* CDC, Atlanta

Figure 5.1: *Childhood blood lead levels considered elevated by the Centre for Disease Control and the Public Health Service, 1965 to 1996*

health considerations. One of the few ways of making sense of that initiative would be if the government wanted to be seen to be doing something about lead, rather than nothing, but without doing what was most important or urgent.

US policy on blood lead targets

The main highlights of US policy on blood lead targets have already been illustrated in Figure 2.2 and are reproduced here as Figure 5.1.

In short, the US government has been setting a blood lead target for children since the 1960s, and it has declined in four steps from 65 μg/dL in the late 1960s to 10 μg/dL in 1991. In that year, the CDC not only reduced its PbB target to 10 μg/dL, but it also recommended a virtually universal screening programme, namely one which would test the blood of all young children in the US except those in communities where large numbers of children had already been tested and shown not to be lead poisoned. The CDC also recommended a programme of primary prevention, under which the authorities would not wait for children to be found to have elevated PbB levels, but would behave proactively to locate and abate the contamination before PbB levels became elevated.[23]

The CDC, in an attempt to provide comprehensive and practical advice, supplemented its baseline target of 10 µg/dL with a six-fold categorization of children with blood lead levels in successive ranges, and specified for each group the kinds of action considered most appropriate. The CDC's advice is given in Table 5.1.

Although it has been the federal government, mainly in the form of the EPA, HUD and the CDC, which has taken the initiative to develop and articulate policy on lead and health, it usually falls to the states and local governments to implement those policies. The federal government, however, has a role to play in trying to ensure that the states fulfil their responsibilities. The CDC has, therefore, asked all 50 states to explain the strategies through which they will be implementing the CDC's and EPA's policies. In a formal exercise reported in 1993, only 48 states replied to the federal government's enquiries, and of those, not all answered every question.[24] Twenty-one states, which corresponds to 44 per cent, had implemented or were planning to implement the CDC's guidelines, while 18 states (38 per cent) planned to phase the guidelines in over several years, but 9 (19 per cent) of the states had no plans whatever to implement the CDC's guidelines.[25]

Because the US government has been so active in recent years, technologies and procedures have been developed to ensure that there are no technical reasons why lead poisoning can not be eliminated. There are a few states suffering from a temporary shortage of laboratory capacity in which to analyse blood samples, but that could easily be overcome. That apart, American institutions have the knowledge and equipment with which the problem can be solved; the same cannot, however, be said of the UK. The main reasons why the CDC's blood lead targets are not being met, and why the EPA's strategy is not being fully implemented, are firstly a lack of funding from Congress and local legislatures, and secondly a lack of interest in some localities.

The policies adopted by the Republican-dominated Congress between 1994 and 1996 have meant that funding for agencies such as the EPA has been sharply reduced and Congress expects the states, counties and cities to fund their own public health and environmental protection programmes. Although cuts in expenditure have occurred in many areas of the EPA's

Table 5.1 *Interpretation of blood lead test results and follow-up activities: class of child based on blood lead concentration*

Class	Blood lead concentration (µg/dL)	Comment
I	<9	A child in Class I is not considered to be lead-poisoned.
IIA	10–14	Many children (or a large proportion of children) with blood lead levels in this range should trigger community-wide childhood lead poisoning activities. Children in this range may need to be rescreened more frequently.
IIB	15–19	A child in Class IIB should receive nutritional and educational interventions and more frequent screening. If the blood lead level persists in this range, environmental investigation and intervention should be done.
III	20–44	A child in Class III should receive environmental evaluation and remediation and a medical evaluation. Such a child may need pharmacologic treatment of lead poisoning.
IV	45–69	A child in Class IV will need both medical and environmental interventions, including chelation therapy.
V	>70	A child with Class V lead poisoning is a medical emergency. Medical and environmental management must begin immediately.

Source: CDC (1991) *Preventing Lead Poisoning in Young Children* October, p3, Table 1.1

activity, programmes concerned with lead pollution have been less severely cut back than many others, but it remains the case that the pace at which human exposure to lead in the US is diminished will depend on the amounts of money devoted to dealing with the problem.

The UK is far less well placed than the US to deal with the problem, because the infrastructure is so poorly developed. However, if the UK government chose to request the co-operation of the US authorities, which they have never done, they could install the technological and organizational systems for markedly less than it cost the US government to set them up. US government officials at the EPA have indicated their readiness to help, if ever their assistance is requested.[26]

One further important difference between the UK and the US is that the US medical profession, unlike their British colleagues, have given the topic considerable attention. For example, in 1995 the American Academy of Pediatrics' Committee on Drugs published a set of guidelines concerned with the use of chelating agents that can be used to treat lead intoxication.[27] The committee recommended that if patients have a PbB level at or below 25 µg/dL then chelation treatment is not appropriate; rather their physical environment should be investigated and action taken directly to reduce exposures. If patients have PbB levels between 25 and 45 µg/dL then the committee recommends aggressive environmental interventions, but even at those PbB levels chelation therapy should not routinely be used. The committee only recommends chelation treatment for patients with PbB levels above 45 µg/dL, and whenever PbB levels exceed 70 µg/dL then the recommendation is that the patient be admitted to hospital and chelation treatment should be administered under clinical management.

Some senior scientists in both the EPA and the CDC have given very serious thought to lowering the official US PbB target for children to a level significantly below 10 µg/dL. Two sets of considerations can be invoked in support of that suggestion. Firstly some recent neuropsychological studies have provided evidence of adverse neurological effects at PbB levels below 10 µg/dL.[28] Secondly there is a case for providing a margin of safety between the lowest level at which adverse effects have been observed and the level down to which children's PbB levels should be lowered. Even though those arguments have been made and debated, EPA and CDC officials have, for the time being, made a tactical decision not to try to lower the PbB target until far more progress has been made towards reaching the 10 µg/dL goal for all children. In the long run, however, it seems unlikely that the current target will remain unchanged once 98 per cent of American children have PbB levels below 10 µg/dL.

UK policy on blood lead targets

The concentrations of lead in the blood of the general public is one of those issues on which the British government has been singularly unforthcoming. The policy of avoiding any discussion of the topic does not however entirely solve the problem. As Figure 5.1 shows, the US authorities have taken the initiative to revise their childhood PbB target downwards on four occasions since the 1960s, and each time they have done so in response to developments in neurotoxicology. In the UK, on the other hand, only one figure has ever been set, primarily in response to pressure from other EC Member States, and when it was set, toxicological considerations played no part in the UK government's decision.

Blood lead targets are not a topic on which the British government has been particularly forthcoming or outspoken. Very few public statements have ever been made and only a handful of policy documents have been issued. The British authorities have never endorsed a blood lead target until they were sure that it had already been met. As a way of stifling debate, controversy and criticism, it has (so far at any rate) been remarkably effective. But that policy has necessitated firstly pretending that the key scientific developments in neurotoxicology over the last 20 years have simply not taken place, and secondly taking less than judicious care of public health.

When, in 1990, Quinn and Sherlock, who were then both British civil servants, discussed the development of UK policy on this topic they implied that, prior to the mid-1960s, the British authorities had no benchmark whatever against which blood lead levels in the general population could have been judged. They suggested, however, that following the publication of two early studies comparing PbB levels amongst mentally retarded children with a group of 'normal' children in 1963 and 1964, a figure of 40 µg/dL '...became the accepted "upper limit of normal".'[29] Quinn and Sherlock don't tell us who 'accepted' it. Maybe some doctors were told during their medical training that PbB levels above 40 µg/dL were undesirable, but there is no evidence that policy makers either endorsed that judgement or even took any notice of it.

In 1977 the EEC promulgated a Directive which obliged the governments of all the Member States to conduct a set of blood lead surveys, and it established three benchmarks by reference to which the results of those studies had to be evaluated,[30] which were, in effect, PbB targets for the general population. The Directive specified that no more than 50 per cent should be above 20µg/dL, no more than 10 per cent should be above 30µg/dL, and no more than 2 per cent should be above 35µg/dL.[31]

As Quinn and Sherlock explain, 'If those levels were exceeded, action was to be taken to trace and reduce the source of exposure. In addition, follow-up investigations were expected if any individual person had a PbB level above 35 µg/dL.'[32]

When, in 1980, the Lawther Committee recommended that, where a child was found to have a PbB level above 35 µg/dL, an investigation should be conducted to identify and reduce their sources of exposure, it was merely reiterating a policy to which the British government, and those of all other EEC Member States, had already agreed three years previously; in those circumstances it was a rather tame proposal. In 1981 the Department of Health went marginally further when it advised that any child with a PbB over 30 µg/dL should be followed up.[33] In 1982 the UK government shifted its position, and set a maximum figure or 'action level' for lead in the blood (PbB) at 25 µg/dL, and until a paper was published in the *British Medical Journal* in November 1994 expert advisors to the British government had never publicly indicated that this action level should be revised downwards.[34]

As discussed in the section on The Strategy of the UK Government, above, in 1982 the Department of the Environment (DoE) advised local authorities, in accordance with the recommendations of the DHSS that '...where a person – *particularly a child* – is confirmed as having a blood-lead level over 25 µg/dL, his or her environment should be investigated for sources of lead and steps taken to reduce exposure.'[35] Local authorities are not however obliged to follow the DoE's advice. The DoE continued: 'Government policy, in the light of medical evidence, is (i) to tackle local environmental "hot spots" where exposure to one source or several is likely to give rise to blood-lead levels over 25 µg/100 ml, and (ii) to seek to reduce exposure generally *as far as is reasonably practicable.*'[36] (emphasis added).

When the British government set an action level at 25 µg/dL they did so by reference to the results of a blood lead survey. Their surveillance work had indicated that the vast majority of the population then had PbB levels below 25 µg/dL, and therefore their endorsement of that figure did not necessitate any further remedial action. There is no evidence that toxicological considerations played any role in the decision to choose 25 µg/dL as an 'action level'. This is therefore an example of the implicit official British rule never to set a target until they are sure that it is already being met.

Throughout the 1980s, however, evidence that lead exerted an adverse neurotoxic effect on children at ever lower levels of exposure continued to emerge, and the pressure group CLEAR continued to press the government to take lead out of petrol. The response of the British authorities to these pressures was the classic one of establishing an investigative committee. Responsibility for the work was assigned to the Medical Research Council (MRC), and a committee was formed chaired by Dame Barbara Clayton, which in the event reported twice, once in 1984 and again in 1988.[37]

These two MRC reviews confined their attention to the evidence of the neurotoxicity of lead, and they passed no comments whatsoever on any other aspects of lead toxicology. The committee did so primarily because that was the brief given to them by the Department of Health and Social Security (DHSS), but the outcome is slightly curious. If the Department of Health had been primarily interested in protecting public health from lead poisoning they would have posed a rather different question namely, 'in which physiological system, and at which lowest level of exposure, are adverse effects first detected?' The concern should have been to establish the lowest level of exposure which would trigger an adverse effect. The question posed by the DHSS was, rather, 'does the evidence on childhood neurotoxicity prove that levels of lead in British children are doing them obvious harm?'

The first MRC report sat resolutely on the fence, listed several of the important studies, and emphasized their method-ological limitations, concluding limply by asserting that if lead was having an adverse neurological effect on British children, that effect was a small one.[38] By 1988 several further studies had been reported, and the second MRC report focused on those recent studies. The committee emphasized many of the method-ological limitations of the studies, but concluded that in the intervening four years the evidence had strengthened. They acknowledged the possibility that '...low level lead exposure has a small negative effect on the performance of children in ability and attainment tests...', and so concluded that '...it would be prudent to continue to reduce the environmental lead to which children are exposed.'[39] That final remark was an acknowledge-ment that the levels of lead to which British children were then being exposed were unacceptably high, although it was couched in language designed not to provoke public anxiety.

The contrast between the 1988 MRC report and the EPA's

1986 report on the *Air Quality Criteria for Lead* was very stark. The MRC provided 18 pages of text, while the EPA's report ran to four thick volumes. The EPA reviewed, if not quite all, at least many of the physiological systems in which lead causes toxic effects, and concluded that low level lead exposure was a serious hazard to public health. The EPA report would have concluded that the blood lead target for children should have been lowered to 10 µg/dL, but other parts of the Reagan administration insisted that that conclusion should be left implicit rather than spelt out. In the event, the CDC endorsed that revised PbB target, but not until 1991. The MRC report did not contain any indication that other aspects of lead toxicology deserved to be considered, and it provided no indication of the PbB levels at which it would be prudent to aim. Moreover, the British government have never subsequently shown any indication that they were even thinking about lowering their PbB target.

The first occasion on which there was any acknowledgement from scientists, whose advice was sought by the UK government, that the long-standing British PbB target of 25 µg/dL needed to be revised downwards came in the *British Medical Journal* in November 1994, ie a good eight years late.[40] The authors, Pocock, Smith and Baghurst concluded that '...a typical doubling of body lead burden (from 10 to 20 µg/dL...blood lead) is associated with a mean deficit in full scale IQ of around 1–2 IQ points...[and]...a threshold below which there is negligible influence cannot currently be determined.'[41] They acknowledged that this result was important for policy-makers, but they failed to articulate the implications.

They qualified their toxicological remarks, moreover, in a way which risked minimizing their impact on policy, because they provided the government with the outlines of an excuse for continued neglect of childhood lead poisoning. They said that the priority which should be devoted to detecting and reducing children's blood lead levels when '....compared with other important social detriments that impede children's development, needs careful consideration.'[42] Of course, careful consideration would be sensible, but given that the British government has been neglecting this matter for too long, and that it has seriously underestimated the magnitude and significance of lead pollution and lead poisoning, they should have been arguing for more rather than less urgency. The French government has, by contrast, already adopted a PbB target of 10 µg/dL, and a major campaign has been launched in Paris to rid old buildings of their lead paint.[43]

Controlling lead in drinking water in the US

Lead is found in drinking water mainly because of corrosion of lead pipes and plumbing fittings. The corrosion is most serious, and the resultant contamination is highest, in areas with relatively acidic water. There are considerable variations between different cities, neighbourhoods, and faucets. Typically, the water supplied by the water company contains very low levels of lead, but it becomes more contaminated the longer it remains in contact with the lead pipes. Consequently, the first flush of drinking water is typically more highly contaminated than the water which subsequently emerges.

Unlike the UK, relatively little lead piping was installed in the US within buildings at the time of their construction, and that practice ceased at around the end of the nineteenth century. Galvanized iron and copper have subsequently been used in preference to lead. Lead has, however, been used to link buildings to the water mains, in what is known as the 'service line'. Millions of lead service lines remain in use in the US. Most American homes, unlike those in the UK, were built with central heating systems. Lead pipes are unsuitable for use in central heating systems because they become very soft when heated, and at or above 140°C they burst.

In 1986 the EPA estimated that approximately 20 per cent of the US population was exposed to levels of over 20 µg/L in first flush water.[44] The EPA estimated, moreover, that (if they made a relatively conservative assumption about the relationship between lead in water and consequent level of blood lead) a steady exposure to 20 µg/L in the drinking water would add between 2.5 and 3.5 µg/dL to a child's blood lead burden. The Agency also estimated that, on average in the early 1990s, drinking water was probably contributing between 1 and 2 µg/dL to the PbB levels of most of the US population.[45]

When, in 1974, under the provisions of the Safe Drinking Water Act, the US set a maximum contamination level for lead in drinking water of 50 µg/L, the government did so primarily by reference to toxicological considerations, rather than considerations of feasibility; by contrast, in the UK considerations of feasibility were invoked, without reference to the toxicological evidence.[46] In 1977, the National Research Council recommended a reduction to 25 µg/L, but the standard was not revised until December 1992.[47]

From 1975 until December 1992 the official American standard specified that the level of lead in tap water should not exceed 50 µg/L. In the light of its concern that the water supply should not contribute to elevated body lead loads, the EPA initially proposed a revised set of rules in 1988 to provide a more restrictive regime, which would have involved lowering the maximum permitted level from 50 µg/L to 15 µg/L.[48] A new set of rules was eventually promulgated by the EPA in June 1991 and these have been enforced since 1 January 1993. These rules specify not just a revised maximum concentration for lead, but also new treatment requirements for drinking water systems.[49]

The regulations require drinking water to be sampled from high risk homes, namely those with lead service lines or with lead solder that has been installed since 1982. The first flush sample must be collected after the water has remained undisturbed for at least six hours. The rule specifies that the 90th percentile of the lead measurements should not exceed the revised action level of 15 µg/L; that is to say, no more than 10 per cent of domestic drinking water samples should contain more than 15 µg/L of lead.

If levels of lead exceed that figure then the utility operator is required to '...optimize corrosion control...' which means that the acidity of the water must be reduced, and other measures such as public education must be provided until the 15 µg/L target (for 90 per cent of the samples) is reached.[50] Water supply companies are, moreover, required to replace lead service lines if corrosion control measures fail to bring lead concentrations (for 90 per cent of the samples) below 15 µg/L.[51] If, after January 1996, the drinking water from a water system continues to exceed that action level, despite corrosion control, then the supplier is required to conduct follow-up monitoring and to start replacing their lead service lines, and they must do that regardless of whether or not the home owners also replace the lead piping within their property. Even though just replacing lead service lines to the boundary of the property will not by itself eliminate all lead contamination, especially if a portion of lead piping remains on the owner's land, it will diminish exposure. The EPA therefore considered that sufficient reason to require a service line replacement programme, without insisting that all lead pipes are removed from the residential property.

In February 1996 the EPA published a set of proposed revisions to the so-called Lead and Copper Rule.[52] Under the new rule, the water service companies are only responsible for the nature and condition of the service pipes up to, but not beyond, the boundaries of private property. The water service company may not, however, refuse to replace the lead service lines unless the home owners also replace theirs. The rule specifies that, when the companies replace the old lead service lines, home owners and occupants must be notified about what is happening and why, and the water service companies must offer assistance. The home owner may then request the company's contractor to replace the lead pipes on their private property when they replace the service pipe, but owners must pay the marginal cost of replacing their section. The home owner will, however, retain the right to decide how little, or how much, of the lead piping on their property to have replaced. This rule is imperfect because the people about whom the EPA is most concerned are those who cannot afford to pay for lead pipe replacement. Some of the cities still served by lead service lines, such as Philadelphia and Boston, are in such poor financial circumstances that they are unable to subsidize a pipe replacement programme, and Congress shows no enthusiasm for making a contribution towards helping poor households in this regard.

In 1986, the EPA estimated that full compliance with the revised drinking water regulations would provide health benefits to the US economy with a monetized value of approximately US$3–4 billion annually, whilst the costs incurred in complying were then estimated at between US$500 and 800 million per year.[53] Calculations of that kind are notoriously imprecise, and assigning monetary values to human health benefits is endlessly contested. One of the problems with the EPA's estimate of the benefits is that they include, in their estimate of the costs of non-compliance, the amount of money that would need to be spent if the children whose development was damaged received remedial educational support sufficient to cancel out the harm which the lead would have inflicted. Since in practice those funds are unlikely to be spent, the EPA might be challenged for having inflated the baseline figure against which to make their comparison. The EPA has a point, of course, namely that reducing lead exposure can be accomplished far more cheaply than providing a corresponding level of remedial educational support. If, therefore, Congress were to invest public resources to enable low income households to participate in lead line replacement programmes, it would in the long run provide a good return. One mechanism through which such programmes could be funded might involve allowing poorer households to deduct from their local property tax liability a proportion of their contribution to the cost of replacing the lead pipes. Those amounts could then be reclaimed by local administrations from the federal government.

Some communities, including for example those in Boston, Massachusetts and San Jose, California, receive their water from quite acidic sources, and the water supply companies have therefore chosen to raise the water's pH by dosing with orthophosphates. Both American and British investigators have found, however, that while adding phosphates can help bring lead concentrations below the 50 µg/L level, it is rarely sufficient to bring it down to 15 µg/L, let alone to 10 µg/L, which is the target recommended by the World Health Organisation and the European Commission.

The argument for replacing lead service pipes is therefore compelling, especially in areas with corrosive water. There is, however, a need to consider the possible impact of an accelerated programme of lead pipe replacement on the current practice of recycling used lead–acid batteries. The viability of the battery recycling system is potentially vulnerable to a sharp decline in the market price for scrap lead. If the rate at which old lead pipes were being replaced rose rapidly, this would tend to drive down the market value of old batteries; it is so much easier to re-use the lead from water pipes than to reclaim the metal from discarded batteries. If lead pipe replacement is not to result in an increase in the unsafe disposal of old batteries, thought will need to be given to ensuring that battery recycling schemes remain viable. One obvious step might be to increase the deposit on the sale of new batteries which is refundable on the return of old ones. At the very least, battery recycling rates should be monitored, and actions taken were they to start to decline.

The use of lead in pipes and solder when connecting drinking water pipes was banned by Congress in the Safe Drinking Water Act of 1986.[54] The EPA has, however, experienced considerable difficulties in enforcing that ban and there is evidence that leaded solder can still be purchased in the US by plumbers and home repairers.[55]

Legislation has been proposed in Congress which includes provisions for setting performance standards on the amounts of lead which may be permitted to leach from plumbing fixtures and fittings. A delay of ten years is envisaged, after which the performance standard would be replaced by a maximum lead content standard. In the light of such concerns, the EPA has been engaged in discussions with industry (through the National Sanitation Foundation or NSF) on the details of a lead leaching standard and the certification of plumbing fittings.[56] Future regulations covering solder and brass plumbing fittings may include a requirement for labels on products which would inform consumers about the restrictions which apply to those products. In 1994 the EPA reached a voluntary agreement which was approved by a majority of NSF members. That agreement obliges manufacturers to submit their products for independent testing, and their equipment will have to meet a maximum lead leaching standard if it is to obtain a seal of approval. The assumption is that equipment not carrying the NSF seal will be far harder to sell.

The EPA has also established standards for bottled water,

and for water supplied from coolers. In 1988 the EPA identified drinking water coolers as a potential source of significant lead exposure, and consequently in 1990 the Consumer Product Safety Commission reached a consent agreement with a major manufacturer of such water coolers to fund their repair and the removal of lead-containing components.[57] In the US, therefore, quite a comprehensive framework of policy is in place, to cover lead in drinking water. There are just two elements which might need to change. Firstly the lead concentration target should be reduced from 15 µg/L to 10 µg/L, to comply with the recommendations of the World Health Organisation, and secondly the resources need to be found to ensure that that target is reached.

Controlling lead in drinking water in the UK

The World Health Organisation (WHO) first recommended a maximum concentration of lead in drinking water in 1958, when a figure of 100 µg/L was set.[58] In their next report, in 1962, the WHO lowered the limit to 50 µg/L, but then raised it again in 1971 in the third edition to 100 µg/L, despite the fact that they had no new scientific information on which to base that decision.[59] No one has ever explained that curious judgement, but it would be reasonable to assume that political considerations played a role.

In 1977 a European Regional WHO Working Party drafted a proposal for a revised lead standard. The vast majority of the members of the working party wanted a maximum of 50 µg/L at the tap without exceptions. There was however a huge row because a small, and primarily British, minority insisted on setting the level at 100 µg/L, though subject to the qualification that

> *Where lead pipes are present, the lead content should not exceed 50 µg/L after flushing. If the sample is taken directly or after flushing and the lead content either frequently or to an appreciable extent exceeds 100 µg/L, suitable measures must be taken to reduce the exposure to lead on the part of the consumer.*[60]

That wording was subsequently incorporated into a 1980 European Community Directive.[61] The Directive set a maximum admissible concentration of lead at 50 µg/L, but crucially, where the distribution system contained lead pipes, the sample was to be taken after flushing. This meant, in effect, that the 50 µg/L limit applied to the water that entered the main, not the water emerging from the consumer's tap, let alone the water actually consumed. At the insistence of the European Commission, and the majority of the Member States, the Directive specified that if lead concentrations exceed 100 µg/L regularly or to an appreciable extent in samples taken at any time, official action should be taken to reduce exposure.

Senior British government officials have acknowledged that the wording of the Directive was '...ambiguous and its interpretation is not straightforward – indeed it exercised the minds of a group of experts for some considerable time.'[62] In the event, it took the Department of the Environment and the National Water Council three years to decide what meaning they would give to those words. They interpreted the phrase to mean that a water company would not be violating the Directive just so long as the average lead concentration in a supply zone did not exceed 50 µg/L from routine samples taken after fully flushing and when taken from the water supply zone as a whole over a period of time.[63] Action should, however, be taken (consisting of an early and detailed investigation) in any water supply zone where more than 2.5 per cent of random daytime samples exceeded 100 µg/L. Action should also be taken to reduce exposure if any individual house showed lead concentrations in excess of 100 µg/L after standing for 30 minutes, in which case nearby houses should also be examined.[64] Those conditions were incorporated into the 1982 guidance notes from the DoE to local authorities explaining who could potentially be eligible for a grant to help meet the cost of lead pipe replacement.[65]

Allowing concentrations of lead as high as 50 µg/L in fully flushed samples was very generous to the water companies, because it applied, in effect, to the water delivered to the earthenware or cast iron main, and not to the water emerging from the kitchen tap. Using that interpretation as a basis for authorizing grants to help with the cost of replacing lead pipes also meant that the presence of lead pipes was never, by itself, sufficient to enable a householder to qualify.

In 1984 the British government relented, mainly under pressure from the European Commission and other Member States, and reinterpreted the Directive to mean that the Maximum Allowable Concentration for lead had to be applied to the water that emerged from the consumer's tap, rather than to the water in the supply pipe. They agreed, moreover, that the limit applied to individual tap samples and not just to the average of numerous samples. In 1989 the Department again shifted policy by announcing a further reinterpretation of the

Directive, saying that from thenceforth it required each and every sample of drinking water to contain no more than 50 µg/L.[66] That evolution of policy has yet to be explained, but a sceptical commentator might think that it consisted of a series of tactical manoeuvres designed to provide the water companies with more time to comply with the Directive, and thereby to facilitate the privatization of the water companies. In 1995 the Council of Ministers and the European Commission agreed to revise the Drinking Water Quality Directive. Under the latest provisions, the concentration of lead in water must fall below 10 µg/L, but that applies to the water delivered to the domestic water system, rather than to the water which people actually drink that emerges from their taps.

The UK Office of Water Regulation (OFWAT) has convened a group which it refers to as its 'National Customer Council' and that group (known as OFWAT NCC) has pointed out that water supplies '.. will not be deemed to have failed the standards if the failure can be attributed to domestic plumbing.'[67] We have not yet reached the stage at which the spirit, rather than merely the letter, of the WHO's standard is accepted throughout the entire EU, and it is the UK which is being particularly recalcitrant. The target intended by the WHO will only be met once the Directive specifies that the level of 10 µg/L is not to be exceeded in any sample of drinking water. Even though a maximum lead concentration of 10 µg/L has been set, water companies are being allowed until 2010 before it needs to be met, with an interim maximum target of 25 µg/L before the start of the twenty-first century. Obviously, at least some governments think that there is no particular urgency.

The OFWAT NCC has also pointed out that '...those organisations which provide food and drink to the public may not be obliged to deal with internal plumbing. [and] There are no obligations on schools, hospitals, restaurants, pubs or similar establishments to remove lead water pipes....[and]...there are no legal obligations on nursery provision (either local authority or private).'[68] Those are important omissions, because without such requirements, it will be harder to ensure that PbB levels fall to, and remain below, 10 µg/dL.

It is ironic, therefore, that the OFWAT NCC suggests that the Directive is over-zealous. The organization suggests that the costs of meeting the standard are excessive in relation to the supposed benefits. That group reached its conclusions, however, by concentrating on narrow, short-term economic considerations, and by ignoring entirely the developments over the last 15 years in lead toxicology. It is not obvious, therefore, that the OFWAT NCC is adequately representing the interests of consumers.

To flush or not to flush, that is the question

The British government's policy on when it is appropriate to flush drinking water to reduce its lead content is not entirely consistent. When water companies are being obliged to test their supplies for lead content, the government has favoured a thorough process of flushing. When it comes to consumers, the government has, however, been far less keen to encourage or authorize a prudent flush of the water supply. In 1982, the Department of the Environment was told by the Water Research Centre (WRC) that flushing the water could provide an appropriate interim means to reduce the public's exposure to lead, pending the complete removal of all lead pipes.[69] The government response to that advice, and to similar comments from the Royal Commission on Environmental Pollution, was hardly enthusiastic or wholehearted. Even though the DoE acknowledged that the margin of safety between actual and desirable levels of expose may have been too small, it then remarked that

> *The main precutionary measure that can be taken to reduce lead intake from plumbosolvent water is to "flush" ie to draw off preferably for some other useful purpose, one pipe volume (usually about a gallon) before drinking water or drawing water for cooking whenever the water has been standing for several hours.*[70]

Instead of instructing the water companies and local authorities to join them in passing the message on to consumers, the DoE said that 'Advice to flush *should not, however, be given indiscriminately* in view of its consequences for the cost of water supply and sewage; it is advisable to consult the local water undertaker before advising to flush.' (emphasis added)[71] That remark implicitly acknowledged that commercial considerations were taking priority over the protection of public health. If the British government only starts to issue advice to flush drinking supplies after domestic water meters have been introduced, that would strengthen the impression that public health had been sacrificed to the commercial interests of the water companies.

The Water Research Centre's 1982 report to the DoE estimated that if households were to flush the lead contamination from their pipes, they might use no more than 100 extra litres a day, which might in turn increase total domestic demand by about 17 per cent. The WRC estimated that the annual cost of that practice would range from 20 pence to a maximum of £1.50 per household, depending on the locality.[72] Those figures imply that the cost of flushing lead out before using the water for drinking or cooking would be slight, while the discussion in the preceding chapters suggests that the health benefits would be significant. Despite the fact that the government has recommended that advice to flush should not be given indiscriminately, some water companies are now providing that advice. Southern Water, for example, issued a leaflet in 1993 which drew their customers' attention to the fact that they can run the tap for about a minute to reduce their exposure to lead, and they did not seem to be worrying about the extra cost.[73]

Treating water for plumbosolvency

A formal obligation on water companies to chemically dose their water supplies to diminish plumbosolvency was introduced in the 1989 Drinking Water Regulations. These specified that, with a few exceptions, wherever there was a risk of the 50 µg/L figure being exceeded, water companies were required to treat the water to reduce its acidity and thereby its plumbosolvency.[74] Compliance with that policy was very cheap for the water companies. In 1982, the WRC estimated that treating drinking water with phosphates to diminish plumbosolvency would entail capital and annual operating costs substantially below the figures of £46m and £7.1m respectively which had previously been suggested.[75]

Water companies in the UK are currently wedded to the use of phosphate dosing to reduce plumbosolvency, because without it they would have been unable to comply with the 50 µg/L standard which they were aiming to reach by 31 December 1995.[76] In May of that year, before the 50 µg/L standard was fully attained, the Commission published a revised draft Water Quality Directive which included a maximum lead concentration of 10 µg/L.[77] The draft Directive envisaged allowing Member States 15 years to reach that target fully, but an intermediate level of 25 µg/L should be achieved within five years.[78] Although phosphate dosing to diminish plumbosolvency should enable most supplies of drinking water to reach the 25 µg/L level, to get below 10 µg/L all lead pipes will have to be replaced.[79] In some areas even that will not be sufficient, in these cases lead-based solders will also need to be removed.[80]

As Friends of the Earth have recently pointed out, 'Universal lead pipe replacement would also decrease the need for phosphate dosing to control plumbosolvency.'[81] This slightly understates the case, because it would entirely eliminate it. The Water Research Centre has pointed out that if lead were not a problem, there would be specific benefits from ceasing to add orthophosphates to water.[82] The addition of orthophosphates encourages the growth of algae where water is stored, promotes the growth of moulds in bathrooms and kitchens, and contributes to the eutrophication of rivers and seas. Making water more alkaline may also be undesirable because it increases the solubility of aluminium, for example from cooking utensils.[83]

Lead pipe replacement policies in the UK

Because there was widespread recognition that significant parts of the British housing stock were in a very poor condition, and that some of the poorest people were in some of the worst housing, a programme was introduced in 1958 to provide what were called 'home improvement grants' to help homeowners upgrade their homes to standards deemed acceptable. In 1982 the rules which governed these grants were modified to include, for the first time, a provision to help meet the costs of lead pipe replacement.[84] The availability of improvement grants was, however, restricted to the poorer homes, and they were designated by reference to what was then called the 'rateable value' of the house which was the officially assessed rental value that was used to determine the householder's liability to local taxation.

Home improvement grants were then available only to owner-occupiers whose property had a rateable value of less than £400 in London or £225 elsewhere in England and Wales; although no such restriction applied in Scotland.[85] The Royal Commission on Environmental Pollution coyly remarked in 1983 that those rules depended on considerations '...not all of which have a direct bearing on public health.'[86] The Commission therefore recommended that '...all owners are given sufficient incentive to expedite pipe replacement in appropriate cases.'[87] Unfortunately the RCEP never provided a definition of what should count as an 'appropriate case', and in any case, the RCEP's recommendation has still not been accepted by the British government.

The National Water Council reported to the RCEP that they had estimated that in 1981 it cost on average £600 to replace the lead water pipes in and to the average dwelling.[88] One-third of the cost was accounted for by the supply pipe, while the other two-thirds covered the costs of replacing the internal plumbing. Replacing a lead-lined tank might on average cost a further £150. In their 1992 report to the Department of the Environment, Chambers and Hitchmough (of the Water Research Centre) estimated that, in total, it would then have cost approximately £2836 million to replace all lead water supply pipes in England and Wales, and about £3101 million to replace internal lead water pipes within the old housing stock.[89] A curious feature of that calculation, however, was that it did not take account of the money which could be obtained from the sale of the scrap lead that would be retrieved. Assuming that approximately 7 million homes would be involved, that each contains an average of 10 m of lead piping, that the average home is reached by a similar length of lead supply pipe, and by making plausible assumptions about the average size of the piping, at prevailing prices, implied that approximately £380 million worth of lead can be retrieved. That figure is a relatively small fraction of the estimated total cost, but since there have been improvements in technology in recent years, the cost estimates are certainly on the high side.

In 1982 the same authors estimated that the average annual rate at which lead pipes were then voluntarily being replaced in England and Wales was about 1.5 per cent.[90] Assuming that rate of replacement remains constant, the last lead water pipes in Britain will be removed in the mid-2040s. There is, therefore, a case for devoting public resources to ensuring that drinking water in the UK meets the standard of 10 µg/L before the middle of the next century.

In February 1996, shortly after the House of Lords Select Committee on the European Communities had published a report warning the government of the dangers of allowing lead pipes to remain in Britain's water supply system, the government took steps to undermine even further the system under which grants were provided to help cover the cost of replacing lead pipes in the homes of poor people.[91] The government introduced the *Housing Grants, Construction and Regeneration Bill* into the House of Lords.[92] The prior arrangements had obliged local authorities to provide grants for the replacement of lead water pipes if the lead levels were consistently too high so as to make the dwelling unfit for human habitation, subject always to a 'means test' by reference to income levels. Under the provisions of the latest legislation, however, all grants from local authorities became 'discretionary', however high the levels of lead in drinking water, however poor the residents, or however elevated their children's PbB levels might be. The policy of the British government, therefore, is to ensure that when the new water quality standard of 10 µg/L is introduced, the costs of compliance should fall, as far as possible, on the private householder and not on the public sector. In circumstances where the resources available to the local authorities are so tightly controlled by the government in Westminster, making the grants discretionary is tantamount to all but abolishing them, while pretending to hold the local authorities responsible for the absence of grants.

Lead pipe replacement and battery recycling

In both the UK and the US, the retrieval and sale of scrap lead piping could have a significant impact on lead battery recycling. The lead from water pipes is relatively clean and pure, and it can easily be recycled. Lead-acid batteries are, however, more difficult, unpleasant and expensive to handle and recycle. There is a risk, therefore, that if no other precautions were taken, a marked growth in the quantities of lead water piping being recycled could undermine the economic viability of the battery recycling system. The consequent danger is that used lead–acid batteries might be discarded into the municipal waste stream, or into ditches or other undesirable locations. It would be unfortunate if the sensible treatment of one lead hazard resulted in the growth of an alternative hazard. That risk could be diminished if the UK were to adopt the kind of system which applies in most states in the US whereby a customer purchasing a replacement battery must either trade in their old battery or pay a deposit that can only be redeemed in exchange for an old battery. A figure of US$10 per battery is not unusual.[93] If the rate at which lead water pipes are removed from US water systems rises sharply, it might become necessary to increase the level of those deposits.

Lead solder in water supplies

In the UK, water companies are empowered to establish local byelaws to prohibit the use of lead solder for drinking water installations. In the south of England, for example, Southern Water has passed such a byelaw, but a ban of this kind is very difficult to enforce, as long as lead solder is still available in the shops.[94] Lead solder continues to be permitted and used, for example, in central heating systems in Britain, although lead solder was banned completely in the Netherlands and Germany in 1977, and subsequently in several states and localities in the US.[95]

In its 1992 environmental policy document *This Common Inheritance – The Second Year Report*, the British government committed itself to banning the use of lead solder in domestic installations but no such ban has been been established.[96] The Department of the Environment subsequently decided, given what it counted as the absence of proof of a direct health risk, and given a desire to avoid 'over-regulation' of industry, to take no further action.[97]

Controlling leaded paint in the US

In 1991, the Centre for Disease Control argued that, in the context of lead policy

> ... *lead-based paint is the source of greatest public health concern. It is the most common cause of high-dose lead exposure. Exposure occurs not only when children ingest chips and flakes of paint (which often contain as much as 50 per cent lead by weight), but also, and probably more commonly, when children ingest lead-based paint-contaminated dust and soil during normal mouthing activities.*[98]

One of the few claims in these complex debates about which there is no dispute is that lead-based paints constitute the most serious single source of lead exposure for the US population. One of the first documented occasions on which a doctor issued a public warning of the dangers of leaded paint for public health occurred in 1904, when Dr Lockart Gibson, a physician in Queensland, Australia, explained why he had concluded that lead paint in the home was responsible for lead poisoning in children.[99]

Since old leaded paint is one of the most important sources of lead for many young Americans, much of the EPA's lead-related effort has been, and remains, focused on addressing the nest of problems which surrounds this complex topic. The EPA has been actively encouraging groups of experts to develop suitable methods for locating lead-based paint, for cleaning-up such paint, and for the disposal of lead-contaminated materials. After the techniques have been developed, the EPA aims to encourage their application and rapid diffusion. For a method of paint abatement to be suitable it must not only be technically efficient, but also cost effective.

One cannot safely scrape or burn the paint as one might with other non-toxic paint residues, because these practices only increase the risk of lead exposure, even if special precautions are taken to seal the premises, and to collect, filter and dispose of the resultant wastes. The cost of lead abatement can sometimes be considerable, although it varies with the thoroughness of the methods. The Federal Department of Housing and Urban Development (HUD) has some responsibility for cleaning up federally-owned housing stock, but meeting the cost of paint abatement in privately owned real estate, especially where it is occupied by poor people, constitutes a very substantial problem to which adequate solutions have yet to be found.

Brugge has recently argued that the companies which historically manufactured and marketed the lead pigments should be held accountable, in proportion to their historical share of the market, for a share of the cost of cleaning up the residues of their previous sales.[100] Brugge claims, moreover, that just five companies produced 90 per cent of the leaded pigments used in American paints.[101] He contends that evidence shows that those companies exerted undue pressure to ensure that the market for lead paint in the US was kept open. He cites, in particular, a record of the Annual Meeting of the Lead Industries Association (LIA) that was held on 5 June 1934, at the Waldorf-Astoria Hotel in New York. The record states that 'an effort was made by the Massachusetts Department of Labor to establish regulations which would have seriously affected the use of white lead in buildings. This subject was discussed by the Secretary [of the LIA] with the State official having the matter in hand and a *satisfactory adjustment procured.*'[102] (emphasis added). The document also states that 'It was particularly important to obtain a hearing and settlement in Massachusetts otherwise we might have been plagued with an extension of similar restrictive painting legislation in other States, affecting the use of white lead.'[103]

In response to the kind of argument which Brugge has been advancing, several members of Congress have tabled legislation for a tax on the current activities of the lead industry, the returns from which would be used to fund the abatement of old paint, but none have received majority support.[104] It is not entirely surprising that proposals along these lines are being vehemently opposed by the lead industry including the LIA. The LIA's view is that leaded paint is a proper focus for regulatory activity, but one for which the lead industry should not have to pay.[105]

Lead-based paint for use in residential contexts was banned in the US by the Consumer Product Safety Commission (CPSC) in 1978.[106] The current position is that responsibility for dealing with residual lead based paint (known in official American circles as 'LBP') in residential units is primarily the responsibility of individual homeowners, although the Department of Housing and Urban Affairs has some responsibility for LBP abatement in public housing.[107]

In 1988 the ATSDR estimated that approximately 3 million

tons of lead remained in paint in US dwellings.[108] Since there were then approximately 14 million children in the US, this corresponded to a staggering figure of some 200 kg of lead per child. Fortunately that statistic fails to accurately represent the nature of the problem. Firstly, it is not just children, but a representative sample of the entire population who inhabit buildings contaminated by LBP. LBP is not uniformly distributed throughout the housing stock, and is almost entirely absent from American homes built after 1978. Some of the lead is not in a state which immediately threatens anyone's health, but potentially it might if it were allowed to deteriorate and contribute significantly to lead levels in ambient air and dust. Leaded paint is often partly, or entirely, covered by one or more subsequent layers of paint. The task for policy-makers is to ensure that as much of that lead as possible is located, evaluated and abated before it reaches a state in which it poses an urgent hazard.

In 1990, the Department of Housing and Urban Development set interim guidelines for the levels above which abatement of lead-based paint in housing should be initiated. They are: 200 $\mu g/ft^2$ for dust on floors, 500 $\mu g/ft^2$ on window sills, and 800 $\mu g/ft^2$ in window wells.[109] The levels of lead contamination which should be reached before abatement can be considered adequate have never been specified, but that issue remains under active consideration.

To try to help cope with the problems of leaded paint, the EPA is both funding research to identify technically and economically effective ways of abatement, and running programmes to train and certify personnel in paint abatement. The US government, especially the EPA and HUD, are investing in training programmes to improve the ability of contractors to do paint-abatement work reliably and safely.

Given that leaded paint on the external surfaces of buildings can contaminate the adjacent soil, and given that children play in those areas, the EPA has conducted a set of trials to investigate the impact on blood lead loads of removing contaminated soils. This research has been conducted in three cities, namely Boston, Baltimore and Cincinnati. The final results should soon be available, but preliminary indications were disappointing, because while it was possible markedly to reduce the lead-content of ambient soils, the consequent reductions in the blood lead loads of the children were rather slight. If this is confirmed in the final report, then the relative priorities of the EPA's strategy may subsequently be modified.

In October 1992, the Residential Lead-Based Paint Hazard Reduction Act was enacted. This provided a legislative framework for a national approach to reducing hazards from lead-based paint exposures primarily from housing. The law requires HUD to provide grants to states to evaluate and reduce hazards from lead in non-federally owned or assisted housing, and to establish guidelines for performing risk assessments, inspection, in-place management and abatement of lead hazards. The EPA is, moreover, mandated to promulgate regulations ensuring that those engaged in abatement activities are trained, and that training programmes are certified. The Agency is also required to establish standards for LBP abatement activities, to promulgate model state programmes detailing how to achieve compliance with the regulations which will govern training and accreditation. The EPA is enjoined, furthermore, to establish a laboratory accreditation programme, establish a clearing house for disseminating information, and to promulgate regulations to cover the disclosure of lead hazards at property sales and transfer. These new responsibilities also include conducting a study on the hazards of renovation and remodelling activities.[110]

Several individual states have also instituted programmes to reduce exposures to lead in residential buildings.[111] Some states are proposing to set restrictions on the ways in which paint abatement can be conducted so as to limit exposures during abatement.[112] For example Texas is proposing a ban on the sand-blasting of painted structures if the paint being blasted contains more than 1 per cent of lead, or if the structure is located near a residential or public area. Minnesota has developed a set of pollution control recommendations for the removal of paint that contains more than 1 per cent lead.[113]

Title X of the Housing and Community Development Act of 1992 is also known as the Residential Lead-Based Paint Hazard Reduction Act (and as Title Ten). It provides a comprehensive legislative framework within which problems arising from lead based paint in housing can be addressed, and in this respect it was a landmark in legislation. Previous statutes had been far more vague, whereas this statute focused not on the mere presence of leaded paint but on the conditions under which the paint poses a hazard to human health. It deals quite differently with homes in the public sector from those which are privately owned. For federally supported housing, specific hazard evaluation and control procedures, activities and standards are prescribed, and it sets specific deadlines and mandates. For private homes built before 1978, the Act firstly requires disclosure of lead-based paint hazards as a precondition for sale, and secondly an obligation to use only certified hazard evaluation and abatement contractors. It imposes quality controls on the lead hazard evaluation and control industry, and requires that contractors gain certification, and seeks to ensure that abatement workers are properly trained and protected.

During the period since October 1992, and under the provisions of that legislation, the heads of the EPA and HUD established a task force to recommend detailed policies on ways of reducing the hazards from lead-based paint, and how the work should be financed. Those efforts culminated in July 1995 in the publication of a report entitled *Putting The Pieces Together*, from the Lead-Based Paint Hazard Reduction and Financing Task Force.[114] One important development which contributed to a recognition that the Task Force had an urgent and important job to do was that in the early and mid-1990s it became increasingly difficult for property owners who rented or leased their property for family homes to be able to obtain so-called general liability insurance to cover them in the event that a child in their property suffered lead poisoning as a consequence of the presence of deteriorated LBP. Contractors were also finding it increasingly difficult to buy suitable insurance cover, without their competence to deal with LBP being officially certified. A new coalition of public and private interests therefore came together to address a problem which some interest groups had previously failed to address.

The Task Force set out to provide benchmark national practical standards for the maintenance of private homes, and for the control of the hazard from LBP, in a way which makes clear the responsibilities of the property owner, and in a way which defines the appropriate ways of responding when children are found to have elevated PbB levels. The Task Force sought to identify ways of ensuring that public finance for hazard control was directed to low income families in economically distressed housing, and it set out to change the liability and insurance arrangements to provide incentives for action to abate hazardous paints and to provide rapid compensation to injured children.[115]

The Task Force has adopted an interesting, and in some ways very un-American tactic, in that it refrained completely from setting any quantitative standards or targets. The Task Force chose instead to describe what it believes to be good practice for local state, county and city policy-makers. It explained the importance of providing public information, training contractors, and ensuring that a wide range of different groups recognize their responsibilities. The Task Force provided lots of useful practical advice. It said, in particular, that old leaded paint should never be removed using dry sanding, open flame burning, power sanding or water blasting; instead it recommended that a surface should be made '...intact by paint stabilization, enclosure, encapsulation or removal.'[116]

The 1992 Housing and Community Development Act also authorized HUD to spend federal resources to help local governments support programmes to deal with hazards from LBP. It established a competitive arrangement under which local adminstrations need to submit reasoned bids in competition for grants to enable them to reduce the PbB hazards in low-income, privately owned housing. Between 1992 and 1996 HUD made grants totalling nearly US$300 million to just under 60 cities and states. Remarkably enough, there was very little disagreement between the Clinton administration and Congress over the need to use federal funds to support programmes to reduce the hazard posed by LBP.

According to Don Ryan, Director of the Washington-based Alliance to End Childhood Lead Poisoning, approximately two-thirds of the US housing stock contains some lead paint, and current estimates suggest that a programme to remove all lead paint would cost in the order of US$500 billion.[117] Resources on that scale are simply not available, and so the Task Force tried to define what they thought was a reasonable and affordable standard of care, which when implemented should provide not lead-free housing, but lead-safe housing. Ryan estimates that compliance with the Task Force's recommendations would on average add less than US$50 a year to domestic maintenance costs. The challenge which remains is to see how far the recommendations are implemented.

In the mid-1990s the US federal government was spending US$65 million on reducing hazards from LBP. Expenditure on that scale is small when compared to the total potential cost, but it has an important role to play in helping to create an organizational and professional infrastructure and enhance the capacity of the system by getting contractors properly trained and certified.

Controlling lead in paint in the UK

The level of lead in paint was first controlled by the British government in 1927, but those early restrictions only applied to paints which were intended for use by industrial workers within factories; paint sold to the general public was not regulated at all until 1989.[118] During the 1950s, the use of lead carbonate and lead sulphate as white pigments was phased out because they were replaced by materials which provided better technical performance, but their use and sale remained legal. In the mid-1980s the UK Paintmakers Association voluntarily agreed to cease using lead additives in decorative paints and varnishes from the start of July 1987.[119] To comply with European standards, British regulations also restricted the use of lead carbonates and lead sulphates in paint except for certain specialist restoration and preservation work. Lead in dry paint film on toys is limited to 0.25 per cent by weight, while the soluble lead content of paints applied to pencils, pens or brushes is limited to 0.025 per cent by weight. All paints and varnishes containing more than 0.15 per cent lead must be so labelled.

The pressure on the British government to set those restrictions has come, once again, from the European Commission and from other Member States of the EC/EU. The government of the Netherlands, for example, prohibited the sale and use of paints containing white lead and lead sulphate in buildings and on boats in 1939. Lead-based paints were twice banned in France, firstly in 1913 and again in 1948.[120] In June 1991 the use of white lead in paints sold to the public was prohibited in France. Germany requires that all paints, glues, degreasers and print colours containing more than 0.1 per cent alkyl lead must be labelled as poisonous, and they must be labelled as harmful if the lead level is between 0.01 per cent and 0.1 per cent. Despite that lack of official action in the UK, the levels of lead in normal commercial paints have declined markedly since the 1920s. That change was a result of technological change in the paint industry, rather than from any official regulatory action.

In 1980 the Lawther Committee said clearly and unambiguously that

> *...steps should be taken in this country [ie the UK] to identify housing areas, schools and play areas where there is a high lead content in existing paintwork. All practicable means should be taken to reduce or eliminate the hazard* as soon as possible *and* research should be conducted *to identify economical and effective methods of doing this.*[121] (emphasis added).

Those recommendations were sound and sensible then, and they remain so today, but they have never been accepted, or followed, by the British government. When the Royal Commission on Environmental Pollution reported in 1983, it pointed out that there were no regulations limiting levels of lead in paint sold to the general public.[122]

From the end of the Second World War until the late 1980s, the British government behaved in a characteristic fashion by discreetly negotiating voluntary agreements with the British industry to encourage it to reduce its reliance on lead-based materials. For example, in 1963 the government concluded an agreement with the Paintmakers Association (PMA) that paints which, when applied and dried, contained more than 1 per cent of their weight as lead (equivalent approximately to 0.5 per cent in the liquid paint) should carry a warning indicating that they should not be used on surfaces accessible to children.[123] In 1974 that agreement was revised and maginally tightened by halving the figures. Under the terms of that agreement, manufacturers were allowed to use the expression 'low lead paint' to label any products which complied with the requirements; and British Standard BS 4310 was introduced to certify that compliance. The substantial majority of British manufacturers were then able to meet the terms of that relatively permissive arrangement, but they could have done a great deal more. The agreement did not apply, in any case, to paints sold by UK companies who chose not to belong to the PMA nor to imported paints

The main consequence of the government's lack of diligence was that the levels of lead in paint in the UK fell less rapidly than they might have done, and less rapidly than occurred in other countries, including the US and several continental European countries. The British government thought that they were helping their domestic industry, but their help was at best short term, and it was negligent in relation to public health. In the longer term the UK paint industry suffered because of reduced exports and competitiveness in international markets, while blood lead levels in Britain have been, and remain, far higher than was necessary or desirable.

Despite extensive research, it has so far proved impossible to locate any evidence that either the British government or the paint industry took any steps to establish the extent to which the advice, issued initially in the mid-1960s, not to use high leaded paints on surfaces accessible to children, was being followed. Indeed, the industry and the government were never very specific about what counted as 'a surface accessible to children'. Occasionally scientists have reported that young children and babies have been observed actually chewing lead-painted wooden surfaces in their homes. As Bryce-Smith has pointed out, lead is almost certainly the sweetest tasting of all known poisons. That is why it is very important that young children do not have access to lead-based paint.

In 1983 the Royal Commission reported preliminary results suggesting that lead levels in the indoor dust of urban homes in Britain then ranged from 39 to 30,060 µg/g, with a mean of 1263 µg/g, averaging over 977 samples.[124] The British government were therefore aware that high levels of lead contamination were occurring as a result of the presence of old leaded paint.

The British government has consistently denied that British children are at risk from this source of lead exposure, meanwhile appropriate survey work has not been taking place.[125] These denials have, however, been repeatedly contradicted by those who were collecting evidence including, for example, the 1983 Royal Commission on Environmental Pollution.[126] The RCEP recommended that 'The government, in collaboration with the paint industry, should establish the quantities of lead-based paints currently sold in the UK.'[127] The government responded with an announcement that the Paintmakers Association of Great Britain had agreed to carry out a survey of its members to establish the quantities of paint made in the UK. The Department of Trade and Industry meanwhile was supposed to compile estimates from those paint manufacturers who were not members of that association.[128] The government pointed out that the Paint Research Association had commissioned research to find alternatives for most uses of lead in paint, indicating as usual that they were not proposing to set any standard until after they could be sure that it had already been met.[129] By the mid-1990s, it was possible to be reasonably confident that leaded paint was not regularly being used to decorate British homes, but the UK government has failed to establish even approximately how many British homes have contaminated paint and dust, or if they have gathered that information, they have never disclosed that fact or published the results.

As Horner pointed out in 1994, 'There is no UK standard for assessing the health hazard from lead in paint'.[130] The UK government has funded some useful but limited research, but that has been confined to a survey of levels of lead in domestic dust. The team of scientists who conducted that research (which included a member of staff of the Department of the Environment) acknowledged that if the action limit for blood lead were to be lowered from its present UK level of 25 µg/dL to even 15 µg/dL, '...many British households would require either some form of remedial action or advice on cleaning procedures.'[131] Given the British government's track record, that is just about sufficient to guarantee that it is extremely reluctant to lower its PbB target.

In 1983, in response to the concerns expressed by the Royal Commission, the British government issued two separate leaflets which contained warnings about the hazard to both adults and children from lead in paintwork.[132] There is no evidence, however, that those leaflets were ever properly distributed, nor that other advisory leaflets on this topic have ever been issued to the general public. As official British advisory leaflets go, they are not too bad. One has very little text and 12 simple illustrations, the other is all text and no pictures. The illustrated leaflet provides the following main items of advice about stripping old paintwork:

- don't burn it off with a blowlamp;
- don't rub it down with dry sandpaper – especially not using a power sander;
- either rub it down using wet waterproof abrasive paper; or
- use a chemical stripper; or
- use a hot air paint stripper
- after stripping, wash your hands thoroughly before eating;
- clean up properly, vacuuming where possible;
- dispose of paint debris safely, for example by putting it in the household waste – do not burn it; and
- repaint with low-lead paint.[133]

The DoE was right to point out that burning off with a blowtorch and dry sanding are the least safe ways of treating old leaded paint, but they are over-optimistic when they recommend chemical paint strippers, hot air guns and wet abrasion.[134] Chemical paint strippers can themselves be very unpleasant and hazardous, unless handled with considerable care. The leaflet fails to point out the importance of not inhaling the fumes released when a hot air gun is used, and it fails to recommend the use of protective clothing such as, at the very least, gloves and masks. Rather than just telling people to 'clean up properly, vacuuming where possible', it should have emphasized the importance of preventing dust and paint flakes from contaminating any area where young children might play.

The text-based leaflet appears to have been aimed at a slightly more serious and well informed audience.[135] It provides a little more information, because it says that hot air tools should be used only well below 500°C. It fails, however, to give any indication as how the user is supposed to tell the temperature at which their tool is operating. It does include advice to remove any carpets before starting work and to keep 'any unnecessary person' out of the room, and to 'wash your skin and hair thoroughly, and change your workclothes before leaving to do anything else – and have those clothes washed well and often.'[136]

A third leaflet, issued in 1990, was aimed at people occupationally exposed to hazards from paint.[137] That leaflet is not, however, specifically about lead, but deals rather with the wide range of different kinds of hazards which exposure to paint, either putting it on or taking it off, can pose, and it focuses primarily on procedures and protective clothing.

In Britain, the general public should have been advised, at the very least, not to strip paint without taking great care to vacuum up all dust and paint flakes, and to dispose of the contents of the vacuum cleaner with considerable care – and not, for example, to put them on a compost heap. The British government has not provided sufficient advice to properly protect public health, it merely recommended the least dangerous of the easily available methods. More importantly the DoE has failed to invest resources into trying to develop more effective, safer and economical methods of paint abatement. There is no evidence that the Department has shown any interest whatever in the lessons learnt by the US government and by the American building trades; a very blind eye is being turned to all the progress which the US has achieved in this area.

More sensible advice might have included recommending that people who are repainting their own homes should prepare lead-contaminated surfaces only lightly and then cover them with good quality lead-free paint. If they choose to have timber surfaces stripped, then the appropriate advice is to have it done professionally, by people who can show that they know how to treat lead-based paint safely. In the US, in contrast, people are officially advised to employ professionals who are trained, licensed and insured. Competent professionals should always seal the areas being stripped so as to isolate them from the rooms in which young children play.

The limitations of the two 1983 leaflets are not important, because the leaflets are virtually unobtainable. I live and work in the East Sussex town of Brighton, where a high proportion of the housing stock was built before1945 and therefore is almost certain to contain LBP. I telephoned each of the major stores in the Brighton area that supply materials and tools for painting and decorating, as well as six of the smaller specialist shops. None of them had copies of the leaflet, and all but two of the shops had never even heard of the leaflet's existence. Brighton is not untypical, and this illustrates the simple fact that the British government is failing the British public, in particular by not following the advice of the Royal Commission.

The contrast with the US could not be more stark. While the US government is investing substantial resources in a comprehensive and workable plan for dealing with old leaded paint, nothing of the sort is taking place in the UK. Official neglect of this important issue is comprehensive, and the British government cannot pretend that it has not been warned.

The primary responsibility for the internal decoration of a house or other building lies, in both the UK and the US, with the person or organization which owns the property. The primary problem in Britain is that the people with the responsibility have little or no relevant knowledge or information, while those with the knowledge are sometimes reluctant to share with property owners either their knowledge or the responsibility for meeting any extra costs which may arise by virtue of the presence of old leaded paint. Once again the contrast between the UK and the US is stark.

In the UK, a responsibility has been laid by central government upon local authorities to ensure that the housing which they own, and other public buildings such as schools, are not contaminated by old leaded paint. In relation to privately owned housing, both homes occupied by their owners and rented properties, the regulatory regime simply empowers environmental health officers, acting on behalf of the local authority, to demand some kind of remedial action, or in extreme cases to condemn a building as unfit for human occupation. They may do so, however, only where there is evidence that occupation of the building is 'prejudicial to health'.[138] The wording of the legislation is, characteristically, vague. None of the legislation, nor the regulations which Ministers have set, provide any practical indication of the conditions under which action should be taken. In principle the flexibility which this permits might be desirable if it meant that environmental health officers would respond rapidly to the emergence of evidence from new toxicological studies. In practice it has meant just the opposite. The local authorities have very few resources and too little information, and so the problem of hazards from leaded paint is almost entirely neglected in the UK.

The Department of the Environment said in 1982 that if a child's blood lead level was found to exceed 25 µg/dL then local authorities were advised (but not obliged) to identify the source of their exposure and to take some remedial action.[139] Issuing that advice has, however, achieved practically nothing. Since there is no systematic programme to monitor the blood lead levels of young children, not even in areas where children live in the kind and age of properties which are likely to contain old leaded paint or where they receive their drinking water through lead pipes, the local authorities only very rarely learn that specific local problems are occurring.

During the late 1980s, a handful of environmental health officers in Britain were able to investigate the presence of old lead-based paint in their local housing stock. Approximately 30 local authorities monitored for old paint and a slightly higher number monitored lead levels in dust or soil.[140] In the 1990s, however, the extent of the monitoring has diminished. As one environmental health officer noted, '...we are simply not resourced to undertake local monitoring if there is not a particular medical problem brought to our attention'.[141] If, however, the British authorities were to lower their blood lead target to 10 µg/dL, and were then to support monitoring activities, it would be possible to gain a far more accurate understanding of the true extent of lead poisoning in the UK; until then guesswork provides the only guide.

6

Reducing the Lead Burden: Past Lessons and Future Progress

Lead is at least as dangerous to the health of young children as the US government believes it to be, and far more hazardous than the British authorities have ever admitted. If young children have blood lead levels above 10 µg/dL then the harm that it will be causing can be detected using currently available experimental methods. If a young child has a blood lead level close to 25µg/dL, a level officially considered acceptable in the UK, its mental, intellectual and educational development would be significantly impaired. We have no biochemical need for lead, and there is no evidence of a threshold below which it exerts no adverse effects.

The US government has an explicit and sensible strategy, though it has yet to demonstrate the political will to implement it fully. The British strategy is unacceptable and unsustainable. It is unsustainable firstly because there is growing pressure from the governments of several other EU Member States, and from the European Commission, to lower the permissible blood lead level at least to 10 µg/dL, and to reduce the maximum concentration of lead in drinking water to 10 µg/L, in accordance with the recommendations of the World Health Organization. Secondly, as the British public gains a greater appreciation of the hazards posed by lead water pipes and lead-based paint, the domestic pressure for change will increase.

The US government has a blood lead screening programme for young children, and it should therefore be able to locate the communities in which excessive exposure to lead is occurring.

Since the UK government has no blood lead screening programme for young children, we do not know how many British children have PbB levels above 10 µg/dL, nor in which communities they are living. Locating those who are over-exposed is, obviously, a first step towards diminishing their exposure.

Following the publication of the results of the most recent nationwide study in the US, Goldman and Carra argued that 'US strategy must begin to focus more than ever on poor, non-white, and inner-city children. We must intensify our efforts to screen these children, identify the sources of lead exposure, and eliminate or reduce these exposures.'[1] The majority of children with too high a body lead burden are certainly living in poor circumstances in inner city areas, but recent work suggests that it would be a mistake to assume that children in rural areas are never exposed to excessive levels of lead.[2] Wherever these children may be, they would benefit from a programme of blood lead screening, and from prompt remedial action wherever excess exposures are revealed.

That task is easier to describe than to complete, however, because in the US, as Goldman and Carra acknowledge: 'Many of these children do not have access to routine medical care....'[3] The failure, between 1993 and 1996, of Congress and the American medical profession to collaborate with the Clinton administration in the reform of the US health care system was a tragically missed opportunity. The British National Health Service has been undermined in recent years, by both the Thatcher and Major administrations, but in some important respects it continues to provide basic health care services that are free to the recipients at the point of delivery. This is one crucial respect in which life for the poor in the US is harder than for those in Britain.

Even without a comprehensive public health care system, American authorities are still capable of organizing and resourcing blood lead screening programmes in communities potentially exposed to excessive levels of lead. If and when the British authorities decide to initiate a childhood blood lead screening programme, the health service will be able to provide a professionally trained and equipped infrastructure, provided that it receives adequate resources.

The Republican-dominated Congress has tended to assign the responsibility for implementing environmental protection and public health protection policies down to the level of the individual states and local administrations. In the name of extending local choice, the congressional authorities have been, in effect, telling local communities that if their health is suffering from excessive lead pollution then they have no one to turn to but themselves. It may be entirely reasonable for Congress to expect local communities to make some contribution to their state-wide and local lead abatement plans, but the federal government ought not to abandon those of its poorest citizens who are not receiving all the help they deserve from their states and local administrations.

The UK government has not committed itself to any detailed policy goals with the exception of trying to ensure that childhood blood levels do not exceed 25 µg/dL and that 'an adequate margin of safety' should be provided. Beyond those vague and out-dated remarks, the British government does not have, and has not had, an explicit strategy to protect public health from lead pollution. It is impossible to estimate, from information in the public domain, how many British children, if any, currently have PbB values in excess of 25 µg/dL. They are likely, however, to be living in communities still receiving their drinking water through lead pipework and in homes with old and deteriorated paintwork.

The British government frequently claims that its public health and environmental policies are based on the best available scientific evidence, but that claim can be no more truthfully made in respect of lead and health than it can in relation to mad cow disease. The British government has a well deserved reputation for believing what its chosen scientists tells it when the advice conforms with ministers' prejudices, and for ignoring the evidence and advice when they do not. The evidence that blood lead levels in children as low as 10 µg/dL can significantly damage their mental performance and development has strengthened to such an extent, and such a strong scientific consensus has developed, that it is becoming increasingly difficult even for British ministers to avert their gaze from unwelcome facts. Other things being equal, the higher a child's blood lead level the poorer will be its ability to benefit from the education with which it is being provided, and if children have blood lead levels above 10 µg/dL, the damage which they would be enduring can be detected using current methods.

The World Health Organization's International Programme on Chemical Safety (IPCS) does not have a reputation for reacting rapidly to new evidence of toxic hazards, especially when the interests of industry are at stake, but even the IPCS has acknowledged that if children's PbB levels rise to about 10 µg/dL then adverse neurological effects will occur – even if they did not quite cast their message in those terms. The failure of the Department of the Environment and the Department of Health to acknowledge or respond to these developments represents a culpable dereliction of their duties. Those departments are not entitled to claim that they were unaware of the scientific developments, since two members of the IPCS Working Party have had long-standing professional links with senior British officials.

The implicit strategy of the British government has been, and remains, one of neglect and irresponsibility. Insofar as standards have been set, they have never been set until they have either already been met, or until the means for showing they are not met have been dismantled. In due course the House of Commons Select Committees for Health and Environment might investigate why the British government did so little for so long.

If the UK government were to start to match the diligence of the American authorities they would start to:

- develop methods to identify geographic 'hot spots';
- implement a lead pollution prevention programme;
- strengthen existing environmental standards;
- develop and transfer cost-effective abatement technology, especially for dealing with leaded paint;
- encourage the availability of environmentally sound recycling;
- develop and implement a public information programme; and
- aggressively enforce environmental standards.

The first step should, therefore, be to establish a proper child-hood blood lead monitoring programme in those parts of the UK known to have older housing and which still receive their drinking water though lead pipework – for example Blackburn and Glasgow. The costs of such a programme would not be prohibitive. Laboratories currently charge between £10 and £20 to provide a PbB measurement from a blood sample. Typically, the higher the price the greater the precision of the results. There are few economies of scale in that kind of analytical work at the moment, but if a sustained and systematic blood lead screening programme were established, there might be scope to halve those laboratory costs. The costs of collecting the samples and delivering them to a suitably equipped and accredited laboratory would be approximately equal to the costs of the analytical work.

It would not be particularly difficult to identify the areas with the highest levels of lead in the drinking water, because all the water companies are obliged to collect water samples and to analyse them for a range of factors including their lead content. That information should be available to, or even already be in the possession of, the Drinking Water Inspectorate. Unfortunately that agency is not yet empowered to use those data in the manner which I am suggesting. More work to monitor lead levels in samples of domestic dust would also be important as part of any systematic attempt to locate children with elevated blood lead levels. In any case, the responsibility for setting the geographical priorities for a blood lead screening programme ought to be assigned to the Department of Health rather than, for example, the Department of the Environment or the water industry.

Given that the World Health Organisation has recommended that water lead concentrations should not exceed 10 µg/L, it would be reasonable to expect that before too long the EPA will also lower its benchmark from 15 µg/L to 10 µg/L. In practice, however, that change may not make a great deal of difference. This is because compliance with the 15 µg/L limit will require the removal and replacement of lead water pipes, and once they have been replaced the residual concentration will normally fall below 10 µg/L.

In both the UK and the US, the water companies are under pressure to remove and replace the lead pipes for which they are responsible. In the long run the main reason why lead pipes will remain in the system will be because they serve the homes of people who are too financially impoverished to meet the costs. It would be reasonable in those circumstances, therefore, to arrange for the more affluent members of society to help their poorer fellow citizens. A straightforward mechanism through which it could be accomplished would be to allow low income households to deduct from their liability to local property taxes the marginal extra costs of having the water companies replace the lead pipes on private property while they are replacing the service lines. The local authorities could then reclaim those amounts from central government. An arrangement along these lines would expedite a worthwhile improvement in public health.

There are both similarities and differences between the challenges which face the UK and the US. Most householders in the US pay for their drinking water supply in proportion to the amount which they use. Water meters are widely used in America but they are comparatively rare in the UK. Most householders in the UK pay for their water by paying a fixed sum which is unrelated to the amount they use, or even to their liability to local property taxes, ie their council tax band. Since the water industry was privatized in 1989, it has been the policy of both government and industry to encourage, and for new homes to require, the introduction of water metering, and charging in proportion to usage. The ostensible reason for favouring the introduction of water meters is that they are supposed to result in a reduction in waste. That argument was undermined in Britain in the mid-1990s when it emerged that the companies' water mains were in such a poor state of disrepair that they were wasting far more clean water from their distribution networks than could be saved by the introduction of metering.

These facts are relevant to the question of the extent to which people with leaded water pipes should be encouraged to flush their supplies before taking a drink or filling their kettles. When in September 1982 the Department of the Environment instructed local authorities not to tell people to flush their water pipes without first checking with the water companies, and perhaps not until water meters had been installed, the government drew attention to the link between the policies on flushing and on metering. If people are only to be encouraged to flush after they have had a meter installed, that will ensure that the victims are obliged to pay both the full cost of flushing, and the full costs of not having previously been advised to flush.

If the government was more concerned about public health than the profitability of the water companies it might have approached the issue in several other ways. If, when the water boards were publicly owned, the government had allowed the companies to follow a proper investment strategy, the mains pipes would never have deteriorated to the point where so much water was being needlessly lost. Once the industry had been privatized, ministers could have set targets for the rate at which leaky mains had to be replaced, or the pace at which wastage rates would need to be reduced. Instead the government were miserly with public resources and extravagant with private ones.

There are at least three options open to a government wanting to make a fresh start. It could either say that consumers only have to pay the full price for water which meets the 10μg/L standard at the tap for any sample, or that consumers can have a sufficient volume of water for free, to provide them with a reasonable margin to ensure that their supply can be adequately flushed. Another option would be to provide resources to contribute to the costs of replacing lead water pipes. Transitional arrangements might involve combinations of those elements.

The main aim of government policy should, however, be to provide both the water companies and the householders with incentives to remove and replace the leaded pipes. The incentives need to be carefully defined because if consumers benefit from extra free supplies or a sharp reduction in prices for receiving their water through leaded pipes, they may be discouraged by the higher cost of receiving uncontaminated supplies.

In any case, consumers who receive their water through lead pipes should be advised by their local authorities to flush their supplies before drinking or cooking with the water if it has been left standing overnight or for several hours. The advice to flush must be given more widely and made more specific, perhaps taking into account the distance from the water main to the kitchen tap. The amount of flushing should be sufficient without being excessive. The decision to provide people with relevant information on how to protect their health should not be inhibited by concerns about the profitability of the water companies. They have been earning high returns on their investments, and the benefits arising from decades of public investment now accrue to the shareholders rather than to the public as a whole. Public policy should be changed to ensure that it becomes in the commercial interests of the water companies to replace their lead pipes, rather than leaving them untouched.

As far as householders are concerned, the government could either allow some or all of the costs of replacing lead pipes on domestic property to be set either against income tax or against their council tax liability. If income tax relief is the only mechanism used to support lead pipe replacement, the help would only be available to those with sufficient income to be liable for that tax, which would mean that the poorer the householder the less likely they would be to benefit. It would therefore be sensible to adopt a different mechanism to help that most vulnerable group.

The water companies in the UK, as in the US, should be obliged to replace their lead service pipes irrespective of whether or not householders also replace theirs. That policy would not necessarily ensure that drinking water always complied with the 10 µg/L standard, but it would contribute significantly to meeting it. The arrangement under which the water companies offer advice and co-operation to the householder is sensible because it enables cost reductions to be achieved whenever householders choose to collaborate with the supplier in pipework replacement. If suitable incentives are provided, the proportion of householders choosing not to join such programmes would be very slight, and should rapidly decline.

Once lead pipes have been fully replaced, it will be possible to cease adding phosphates to the water supply, and that too should provide extra benefits, in terms of producing conditions less favourable to the growth of algae and moulds, and the eutrophication of rivers and seas.

The 1995 EU Directive on Drinking Water Quality allows the water companies until the year 2000 before they have to provide supplies with lead concentrations below 25 µg/L and until 2010 before they have to comply fully with the target of 10 µg/L. One of the shortcomings of that legislation is that it is, in effect, predicated on the assumption that all the adverse toxicological information which can emerge on lead has already done so. The history of the last 25 years strongly suggests, however, that those targets will be completely out of date before they are reached.

In the UK there are approximately 10 million homes which receive their drinking water through some portion of lead piping. Given that recently, on average, approximately 100,000 homes a year have had those lead pipes replaced, the requirements of the latest Directive could be reached by increasing that number to about 700,000 homes per year. The high rates of return achieved in recent years by the privatized British water companies suggest that they could easily afford to raise that replacement rate. There are no technical or commercial reasons why the government should not require them to comply with the 10µg/L standard in a shorter period than the 15 years allowed by the Directive.

In the interim several steps could usefully be taken. People should be warned not to fill their kettles or saucepans from their hot water taps, and the sale of leaded solder should be more carefully controlled. Professional plumbers know that they should never use lead-based solders in drinking water systems, but their compliance is being taken for granted, and as General Eisenhower is reputed to have said, 'the uninspected deteriorates'. The British consumer organizations ought, moreover, to ask some very searching questions about how adequately their interests are being represented by the OFWAT NCC. When an officially appointed body officially appoints those who speak on behalf of the customer, one is entitled to ask how adequate are the lines of accountability?

Turning finally to policies on lead-based paint (LBP), the evidence provided in the previous chapters indicates that the American authorities have a well thought out and detailed set of policies to reduce substantially the hazards posed to children and the general public from LBP. They appreciate that it is difficult to bring up a young child in a house contaminated with lead-based paint in a deteriorated condition without that harming the child's mental development. The main difficulty is that the resources needed to implement the policy of the US government are not yet being provided. Progress is being made, but at a rather modest rate. The people who stand to benefit most from increased levels of investment are almost certainly those least well placed to influence the policy process. It is therefore fortunate that there are those, like Don Ryan and his colleagues in the Alliance to End Childhood Lead Poisoning, civil servants in the EPA, FDA and CDC (amongst others) and academic scientists too numerous to mention, who have been trying hard to raise the profile of this issue on the public agenda.

The contrast with the position on LBP in the UK could hardly be greater. The only policy in place in the UK is one of neglect. It is hard to imagine a blinder eye being turned.

I have devoted over 20 years of my career to exploring the toxic hazards for which food additives might be responsible. I vividly recall a conversation with a senior scientist at the British food and chemical industries' research centre in Carshalton, when in the late 1980s he exclaimed: 'Why don't you stop wasting your time on food additives and concentrate instead on something which is really poisonous, like lead?'

Without in any way conceding his claim that food additives are invariably innocuous, he is right that lead poisoning is taking place, it is demonstrably damaging many children, and it could easily (though not necessarily cheaply) be avoided.

Given therefore that the US has done so much on LBP and the UK has done practically nothing, the obvious strategy for a British government determined to remedy its history of negligence would be to establish a working relationship with the relevant parts of the EPA and HUD, and start learning from the experience of the Americans.

A high powered team of British civil servants, public health authorities, and representatives of the building and decorative trade should be sent on a fact-finding mission to the US. On their return, they should prepare a report outlining how a strategy to reduce the hazards from LBP should be implemented in the UK. Their recommendations will necessarily involve proposals concerning the provision of public information to householders and to those who choose to redecorate their own homes, as well as to professional tradespeople.

The British government needs to stop pretending that no problem exists, and start by examining, and reporting on, the extent and severity of deteriorated LBP in the British housing stock. They should consult with a wide range of interested parties, and then publish a strategic plan for reducing exposure to hazardous LBP, and should indicate the resources which they propose to devote to the problem, once its magnitude has been established. As long ago as 1980, the Lawther Committee recommended that 'There should be a programme for the detection of lead in paint coatings accessible to children in areas where a high incidence of old lead paint surfaces may be suspected, such as old inner city residential areas.'[4] The failure to implement that recommendation, and follow through its implications, can no longer be excused.

The London-based Lead Development Association has claimed that upwards of 90 per cent of old lead-based batteries get recycled in Britain, while a so-far unpublished study by a singularly well informed industrial consultant suggested that the true recycling rate for batteries in the UK is closer to 50 per cent. If that latter estimate is correct, it would mean that thousands of old lead–acid batteries are either being disposed of inappropriately or remain stored unsafely in people's homes, garages and gardens. As and when the rate of replacement for old leaded water pipes accelerates, as it is bound to do under the provisions of the 1995 Directive, the market price for scrap batteries may well decline, and decline sharply. That could result in a sharp decline in the rate at which old batteries are recycled, whatever the true current baseline figure may be. Once again, American experience shows that a simple deposit system can make a very useful contribution to raising and maintaining recycling rates. A scheme of that sort will become essential in Britain in the near future.

It is often all too easy, either from the eastern or the western side of the Atlantic Ocean, to assume that things are better across the other side of (what the British sometimes refer to as) the 'pond'. The evidence outlined in this book shows that in most, but not quite all, respects hazards to public health from lead exposure are being addressed more effectively in the US than in the UK. Not everything in the US is perfect, but the state of affairs in the UK is very significantly worse. The US has a policy framework and a strategy in place – all it needs now are the political will and resources. The UK, however, has the advantage of a National Health Service, and if the institutions and personnel in the NHS were suitably resourced and mobilized in a childhood blood lead screening programme, and if the British government showed a willingness to learn from the American authorities, considerable progress could be made, providing significant improvements in public health.

Endnotes

Notes to Chapter 1

1 Alliance to End Childhood Lead Poisoning (1991), *Preventing Childhood Lead Poisoning: Conference Notebook* The First Comprehensive National Conference, Washington DC, October, pA–2

2 CDC (1991) *Preventing Lead Poisoning in Young Children: A Statement by the Centre for Disease Control* CDC, Atlanta

3 op cit p1

4 Lin-Fu, J S (1992) 'Modern history of lead poisoning: a century of discovery and rediscovery' in Needleman, H L (ed) *Human Lead Exposure* CRC Press, Boca Raton, Florida, p24

5 US EPA (1986), *Air Quality Criteria for Lead, Volumes I to IV* June, document numbers EPA–600/8–83/028aF to EPA–600/8–83/028dF, Washington DC

6 See eg Quinn, M J and Sherlock, J C (1990) 'Correspondence between UK "action levels" for lead in blood and in water' *Food Additives and Contaminants* vol 7, no 3, pp402–3

7 CDC (1991) op cit

Notes to Chapter 2

1 At least since Carlyle first characterized economics in those terms: Carlyle, T (1850) *Latter-Day Pamphlets* No 1, *The Present Time* Chapman

2 See, however, Millstone, E (1986) *Food Additives* Penguin Books, London; and from a rather different perspective Efron, E (1984) *The Apocalyptics*, Simon & Schuster, New York, Parts III and IV

3   Arnold, D L, Krewski, D and Monroe, I C (1983) 'Saccharin: A
     Toxicological and Historical Perspective' *Toxicology* July–August
     vol 27, Part 3–4, pp179–256

4   Orfila, M P (1817) *A General System of Toxicology* M Carey &
     Sons, Philadelphia; quoted by Goyer, R A (1991) in 'Toxic Effects
     of Metals' in Amdur, M O, Doull, J & Klaassen, C D (eds)
     Casarett and Doull's Toxicology: The Basic Science of Poisons
     Pergamon, p639

5   Department of Health and Social Security (1980) *Lead and
     Health* (also known as *The Lawther Report*) HMSO, p7, para 2

6   See EPA (1986) *Air Quality Criteria for Lead* vol 1, pI–134

7   *Australian/International Meeting on Non-Occupational Exposure to
     Lead: Meeting Summary* Melbourne, October 1992, p3 of draft
     minutes

8   Evidence emerged in 1995 indicating that lead levels in the shin
     bone can be measured using an x-ray fluorescence technique.
     But this method is expensive to apply, and its clinical utility
     remains to be demonstrated. Roels, H et al (1995) 'Time-
     integrated blood lead concentrations a valid surrogate for
     estimating the cumulative lead dose assessed by tibial lead
     measurement' *Environmental Research* vol 69, pp75–82

9   Fulton, M (1989) 'Lead exposure and child development – some
     methodological issues' *Heavy Metals in the Environment* vol 1,
     p95

10   Cory-Slechta, D (1990) 'Exposure duration modifies the effects of
     low-level lead on fixed interval performance' *NeuroToxicology* vol
     11, pp427–42; quoted in Bellinger, D (1995) 'Interpreting the
     literature of lead and child development: the neglected role of the
     "experimental system"' *Neurotoxicology and Teratology* vol 17,
     no 3, p203 fn 24

11   A few contributions to an extensive and complex debate include:
     J Julian Chisholm Jr (May 1976) Editorial: 'Current Status of
     Lead Exposure and Poisoning in Children' *Southern Medical
     Journal*; Needleman et al (1979) *New England Journal of
     Medicine* p689; Lin-Fu, J (March 29 1979) *New England Journal
     of Medicine* p732; Yule, W and Rutter, M (1985) 'Effects of lead
     on children's behavior and cognitive performance: a critical
     review' in Kathryn R Mahaffey (ed) *Dietary and environmental
     lead: human health effects* Elsevier, Amsterdam; Smith, M et al
     (eds) (1989) *Lead Exposure and Child Development* Kluwer
     Academic Press, Dordrecht, Ch 1.1, p4 ; Fulton, M (1989) 'Lead
     exposure and child development – some methodological issues'
     Heavy Metals in the Environment vol 1, p95

12   Cf Needleman and Gatsonis, 1990, p677

13   J Julian Chisholm Jr (May 1976) Editorial: 'Current Status of
     Lead Exposure and Poisoning in Children' *Southern Medical
     Journal* pp529–31

14 Yule, W and Rutter, M (1985) 'Effects of lead on children's behavior and cognitive performance: a critical review' in Kathryn R Mahaffey (ed) (1985) *Dietary and environmental lead: human health effects* Elsevier, Amsterdam, p215,

15 In 1982, for example, Mahaffey and her colleagues reported that the imprecision in estimating blood lead levels was between 12 and 15 per cent, expressing the standard deviation as a percentage of the average of multiple measurements. K R Mahaffey et al (1982) 'National estimates of blood levels: United States, 1976–1980: associated with selected demographic and socioeconomic factors' *New England Journal of Medicine* vol 307, p574

16 *Australian/International Meeting on Non-Occupational Exposure to Lead: Meeting Summary* Melbourne, October 1992 - meeting in support of the World Health Organisation's International Programme on Chemical Safety's review of Lead; cf World Health Organisation's International Programme on Chemical Safety (1995) *Inorganic Lead* Environmental Health Criteria 165, Geneva, Sec 9.4.1

17 Kramer, M S (1979) 'Lead Levels and Children's Psychological Performance' *New England Journal of Medicine* vol 301, no 3, p161

18 eg Needleman, N L et al (1979) 'Deficits in psychologic and classroom performance of children with elevated dentine lead levels' *New England Journal of Medicine* vol 300, no 13, pp689–95; Yule, W and Rutter, M (1985) 'Effects of lead on children's behavior and cognitive performance: a critical review' in Kathryn R Mahaffey (ed), *Dietary and environmental lead: human health effects* Elsevier, Amsterdam, p233

19 op cit, p222

20 Smith, M, Delves, T, Lansdown, R, Clayton, B E and Graham, P (1983) 'The effects of lead exposure on urban children: the Institute of Child Health/Southampton Study' *Developmental Medicine and Child Neurology* vol 25, suppl 47, pp1–54

21 Fulton, M (1989) 'Lead exposure and child development - some methodological issues' *Heavy Metals in the Environment* vol 1, p95; cf Fergusson, D M et al (1988) 'A Longitudinal Study of Dentine Lead Levels, Intelligence, School Performance and Behaviour: Part I. Dentine Lead Levels and Exposure to Environmental Risk Factors' *Journal of Child Psychology and Psychiatry* vol 29, no 6, p782

22 See eg EPA, *Air Quality Criteria for Lead* (1986); ILZRO, 1991, Appendix I, p6; Winneke et al, 1990, p554

23 Ward, N & Stovell, A (Easter 1996) 'Trace Element Hair Analysis for the Diagnosis of Human Health Status' *Foresight Newsletter*, Godalming, pp20–22

24 US Court of Appeals for the District of Columbia Circuit, Case numbers 91–1338 and 91–1343, decided December 6, 1994

25 Department of the Environment (1983) *European Community Screening Programme for Lead: United Kingdom Results for 1982* Pollution Report No 18 and Quinn M J (1985) Factors affecting blood lead concentrations in the UK *International Journal of Epidemiology*, 14, 420–431; also see Quinn and Sherlock (1990) p390

26 Department of the Environment and Welsh Office (1982) *Lead in the Environment*, Circular 22/82 London HMSO

27 *Pers. Comm.* between J Russell and Dr Andrew Wadge, Principle Scientific Officer, UK Department of Health, 20th December 1993; and between E Millstone and M Waring, June 1997

28 ILZRO, *Comments on Health and Toxicology Sections of the OECD Lead Document*, Aug 1991, p1

29 op cit p7

30 Ernhart, C B (1992) 'A Critical Review of Low-Level Prenatal Lead Exposure in the Human: 2. Effects on the Developing Child' *Reproductive Toxicology* vol 6, p37

31 *The Australian/International Meeting on Non-Occupational Exposure to Lead: Meeting Summary* Melbourne, October 1992, p6

32 ATSDR (1988) *The Nature and Extent of Lead Poisoning in Children in the US: A Report to Congress* July; quoted by Wong et al (1992) p288

33 ILZRO (1991) *Comments on Health and Toxicology Sections of the OECD Lead Document* August 1991 p2

34 Even if techniques of chemical analysis were to improve markedly, many of the methodological difficulties would remain. As Flegal and Smith point out, 'Future studies to assess the threshold concentrations of subclinical lead toxicity will be limited by the difficulties in establishing suitable control populations.' Flegal, A R and Smith, D R (1992) 'Current Needs for Increased Accuracy and Precision in Measurements of Low Levels of Lead in Blood' *Environmental Research* vol 58, pp125–133

35 Goyer, R (1991) 'Toxic Effects of Metals' in Amdur, M O, Doull, J & Klaassen, C D (eds) *Casarett and Doull's Toxicology (The Basic Science of Poisons)* Pergamon, p639

36 Department of Health and Social Security (1980) *Lead and Health* HMSO, also known as *The Lawther Report*; cf Yule, W and Rutter, M (1985) 'Effects of lead on children's behaviour and cognitive performance: a critical review' in Kathryn R Mahaffey (ed) *Dietary and environmental lead: human health effects* Elsevier, Amsterdam, p211

37 ILZRO (1991) *Comments on Health and Toxicology Sections of the OECD Lead Document* August, p12 (cf Appendix I, p23)

38 Schwartz, K (1975) 'Potential Essentiality of Lead' *Ark Rada Toksikol* vol 26 (suppl) pp13–28; cited by ILZRO (1991) p12

39 Schwartz, K (1975) 'Potential Essentiality of Lead' *Ark Rada Toksikol* vol 26 (suppl) pp13–28; cited by ILZRO (1991) p12 (cf Appendix I, p23)

Notes to Chapter 3

1   Schwartz, J (1994) 'Societal benefits of reducing lead exposure' *Environmental Research* vol 66, p121

2   US EPA (1986) *Air Quality Criteria for Lead* vol IV, p13–31

3   World Health Organisation's International Programme on Chemical Safety, *Inorganic Lead*, Environmental Health Criteria 165, Geneva, 1995, pp31–32

4   CDC (Oct 1991) *Preventing Lead Poisoning in Young Children: A Statement by the Centre for Disease Control*, CDC, Atlanta, Fig 2.1, p8

5   Sherlock, J C and Quinn, M J (1986) 'Relationship between blood lead concentrations and dietary lead intake in infants: the Glasgow duplicate diet study 1979–1980' *Food Additives and Contaminants* vol 3, no 2, p167

6   Goyer, Robert A (1991) 'Toxic Effects of Metals' in M O Amdur, J Doull and C D Klaassen (eds) *Casarett and Doull's Toxicology (The Basic Science of Poisons)* Pergamon, p641; cf ILZRO (1991) *Comments on Health and Toxicology Sections of the OECD Lead Document* August, p4

7   Thomas, H M and Blackfan, A D (1985) *J Dis Child* 1914 vol 8, pp377–80; cited by W Yule and M Rutter, 'Effects of lead on children's behaviour and cognitive performance: a critical review' in Kathryn R Mahaffey (ed) *Dietary and environmental lead: human health effects* Elsevier, Amsterdam, p212

8   Yule and Rutter (1985) p212

9   Smith, M A, Grant, L D and Sors, A I (eds) (1989) *Lead Exposure and Child Development* Kluwer Academic Press, Dordrecht, p488

10   Yule, W and Rutter, M (1985) 'Effects of lead on children's behavior and cognitive performance: a critical review' in Kathryn R Mahaffey (ed) *Dietary and environmental lead: human health effects* Elsevier, Amsterdam, p223

11   See eg Butcher, H J (1970) *Human Intelligence: Its Nature and Assessment* Methuen & Co, London, see esp Chapter IX 'A selective survey of intelligence tests' and Chapter X 'Social and cultural influences'

12   Bellinger, D (1995) 'Interpreting the literature of lead and child development: the neglected role of the "experimental system"' *Neurotoxicology and Teratology* vol 17, No 3, p203

13   ibid

14   See eg Ryan, J (1972) 'IQ - The Illusion of Objectivity' in K Richardson and D Spears (eds) *Race, Culture and Intelligence* Penguin Books, chapter 2, p36 ff

15   Fulton, M (1989) 'Lead exposure and child development – some methodological issues' *Heavy Metals in the Environment* vol 1, p96

16   Medical Research Council (1988) *The Neuropsychological Effects of Lead in Children: A Review of Recent Research 1984-1988* paragraph 5.13, p19

17   *Australian/International Meeting on Non-Occupational Exposure to Lead: Meeting Summary* Melbourne, October 1992, p4

18   Baghurst, P A (1995) 'Getting the lead out...' *Neurotoxicology and Teratology* vol 17, no 3, p213; The term 'standard deviation' is a technical term coined by statisticians and is used as an indicator to represent the variation within a population; the smaller the standard deviation the more closely clustered together are the members of the population, while the larger the standard deviation, the more people there are towards the two extremes of the distribution. Strictly speaking, the standard deviation is the positive square root of the mean of the squares of the differences between each observation and the average of all the observations

19   S Rose 'Scientific Racism and Ideology: The IQ Racket' *The Political Economy of Science*, p119

20   Ryan, J (1972) p54

21   *Australian/International Meeting on Non-Occupational Exposure to Lead: Statistical Methods Group Minutes* (October 1992) Melbourne, p1

22   Pocock, S J and Ashby, D (1985) 'Environmental lead and children's intelligence: a review of recent epidemiological studies' *The Statistician* vol 34, pp33-4

23   US EPA (August 1990) *Air Quality Criteria for Lead: Supplement to the 1986 Addendum* pp52-3

24   Medical Research Council (1984) *The Neuropsychological Effects of Lead in Children: A Review of Recent Research 1979-1983*, p5

25   Yule, W and Rutter, M (1985) 'Effects of Lead on Children's Behavior and Cognitive Performance' in Kathryn R Mahaffey (ed) *Dietary and environmental lead: human health effects* Elsevier, Amsterdam

26   Yule and Rutter op cit, p213

27   Yule and Rutter op cit, p213 and p216

28   Yule and Rutter op cit, p213

29   As Yule and Rutter pointed out in 1985, op cit, p214

30   Ruff, H A (1993) 'Declining blood lead levels and cognitive changes in moderately lead poisoned children' *Journal of the American Medical Association* 7 April, vol 269, no 13, pp1641-46

31   Its full name is calcium disodium ethylene diamine tetra-acetate

32   Ruff, H A (1996) 'Relationship among blood lead levels, iron deficiency, and cognitive development in two year-old children' *Environmental Health Perspectives* February vol 104, no 2, p180

33   cf Yule and Rutter (1985), p214

34   See eg Lansdown, R G et al (1974) 'Blood-lead levels, behaviour, and intelligence: a population study' *Lancet* 30 March pp538-41;

Landrigan, P J et al (1975) 'Neuropsychological dysfunction in children with chronic low-level lead absorption' *Lancet* March 29, pp708–12; Yule, W et al (1981) 'The Relationship Between Blood Lead Concentrations, Intelligence And Attainment in a School Population: a Pilot Study' *Developmental Medicine and Child Neurology* vol 23, pp567–76; Lansdown, R et al (1983) 'Blood Lead, Intelligence, Attainment and Behaviour in School Children: Overview of a Pilot Study', in M Rutter and R Russell Jones (eds) *Lead versus Health* Wiley, pp267–96

35 Lansdown, R G et al (1974) p541

36 Yule and Rutter (1985) p214

37 Hatzakis, A et al (1988) 'Psychometric Intelligence Deficits in Lead-exposed Children' in Smith, M A et al (eds) *Lead Exposure and Child Development* Kluwer Academic Publishers, Dordrecht, p211–23

38 Hatzakis op cit, p212

39 Hatzakis op cit, pp219–221

40 Needleman, N and Gatsonis, C (1990) 'Low-Level Lead Exposure and the IQ of Children, A Meta-analysis of Modern Studies' *Journal of the American Medical Association* 2 February, vol 263, no 5, pp673–78

41 Needleman, H L et al (1979) 'Deficits in psychologic and classroom performance of children with elevated dentine lead levels' *The New England Journal of Medicine* March 29, vol 300, no 13, pp689–95

42 Needleman et al op cit, p692

43 Yule and Rutter (1985) pp234–5

44 Kramer, M S (1979) 'Lead Levels and Children's Psychological Performance' *New England Journal of Medicine* 19 July, vol 301, no 3, p161

45 Hall, D M (1979) 'Lead Levels and Children's Psychological Performance' *New England Journal of Medicine* July 19, vol 301, no 3, p161; Cole, J 'Lead Levels and Children's Psychological Performance' op cit, p162

46 Coplan, J (1979) 'Lead levels and children's psychological performance' *New England Journal of Medicine* 19 July, vol 301, no 3, p162; Lynam, D R 'Lead Levels and Children's Psychological Performance' *New England Journal of Medicine* vol 301, no 3, pp162–3

47 Needleman (1979) p163

48 Complaints were also made concerning the adequacy and reliability of some early work by Perino and Ernhart. See J Perino and C Ernhart, (1974) 'The relation of sub clinical lead level to cognitive and sensorimotor impairment in black pre-schoolers' *Journal of Learning Disorders* vol 7 pp772–82

49 US EPA (1983) Appendix 12-C [of EPA 1986 Lead Criteria

document], *Independent Peer Review of Selected Studies Concerning Neurobehavioral Effects of Lead Exposures in Nominally Asymptomatic Children* 14 November

50 US EPA op cit, pvi

51 Palaca, J (1991) 'Get-the-Lead-Out Guru Challenged' *Science* vol 253, 23 August, p843

52 US EPA (1986) *Air Quality Criteria for Lead* Volumes I to IV, documents numbered EPA–600/8–83/028a-dF; CDC (1991) *Preventing Lead Poisoning in Young Children: A Statement by the Centre for Disease Control* CDC, Atlanta, October; ATSDR, *Toxicological Profile for Lead*, Draft for Public Comment, October

53 Ernhart, C B and Scarr, S (1992) 'Lead Study Challenge' letter in *Science* 14 February, vol 255, p783

54 Hilts, P J (1992) 'Hearing Is Held on Lead-Poison Data' *New York Times* 15 April pD28

55 Wishart, A (1992) 'Lead campaigner cleared of fraud' *New Scientist*, 30 May, p7

56 Palaca, J (1992) 'Panel Clears Needleman of Misconduct' *Science* vol 256, 5 June, p1389

57 Taylor, R (1992) 'Pitt's Fuzzy Verdict in Needleman Case' *Journal of NIH Research* December vol 4, p44

58 ibid

59 Palaca, J (1991) 'Get-the-Lead-Out Guru Challenged' *Science* vol 253, 23 August, p842

60 Smith, M, Delves, T, Lansdown, R, Clayton, B E and Graham, P (1983) 'The effects of lead exposure on urban children: The Institute of Child Health/Southampton Study' *Developmental Medicine and Child Neurology* vol 25, Suppl 47, pp1–47

61 See eg Yule and Rutter (1985) p240; Bellinger, D et al (1989) 'Low-level Lead Exposure, Social Class and Infant Development' *Neurotoxicology and Teratology* vol 10, pp497–503; Bellinger, D et al (1989) 'Lead, IQ and Social Class' *International Journal of Epidemiology* vol 18, no 1, pp180–185

62 Yule and Rutter (1985) p239

63 Medical Research Council (1984) *The Neuropsychological Effects of Lead in Children: A Review of Recent Research 1979–1983* p15, para 4 5

64 Yule and Rutter (1985) p241

65 See: Fulton, M et al (1987) 'Influence of blood lead on the ability and attainment of children in Edinburgh' *Lancet* 30 May, pp1221–25; Fulton, M et al (1989) 'Blood lead, tooth lead and child development in Edinburgh' *Heavy Metals in the Environment* vol 1, pp68–71; Fulton, M (1989) 'Lead exposure and child development – some methodological issues' *Heavy Metals in the Environment* vol 1, pp94–102; and Fulton, M (1989) 'Exposure to low levels of lead in the environment: assessing its

effect on the mental development of children' *Proceedings of the Royal College of Physicians of Edinburgh* vol 19, pp15–22

66   Fulton, M et al (1987) 'Influence of blood lead on the ability and attainment of children in Edinburgh' *Lancet* 30 May, p1221

67   Fulton et al op cit, p1222

68   Fulton et al op cit, p1223

69   Medical Research Council (1988) p8, para 3.24

70   Medical Research Council (1988) p8, para 3.24

71   Fergusson, D M et al (1988) 'A Longitudinal Study of Dentine Lead Levels, Intelligence, School Performance and Behaviour: Part I. Dentine Lead Levels and Exposure to Environmental Risk Factors' *Journal of Child Psychology and Psychiatry* vol 29, no 6, pp781–92; Fergusson, D M et al (1988) 'A Longitudinal Study of Dentine Lead Levels, Intelligence, School Performance and Behaviour: Part II. Dentine Lead and Cognitive Ability' *Journal of Child Psychology and Psychiatry* vol 29, no 6, pp793–809; Fergusson, D M et al (1988) 'A Longitudinal Study of Dentine Lead Levels, Intelligence, School Performance and Behaviour: Part III. Dentine Lead Levels and Attention/Activity' *Journal of Child Psychology and Psychiatry* vol 29, no 6, pp811–24

72   Fergusson et al op cit, p797

73   Fergusson et al op cit, pp799–800

74   Fergusson et al op cit, p805

75   Needleman, N and Gatsonis, C (1990) 'Low-Level Lead Exposure and the IQ of Children, A Meta-analysis of Modern Studies' *Journal of the American Medical Association* 2 February, vol 263, no 5, p673

76   op cit pp673–78

77   Winneke, G et al (1990) 'Results from the European Multicentre Study on Lead Neurotoxicity in Children: Implications for a Risk Assessment' *Neurotoxicology and Teratology* vol 12, pp553–9

78   World Health Organisation's International Programme on Chemical Safety (1995) *Inorganic Lead* Environmental Health Criteria 165, Geneva

79   See eg Glass, G V, McGaw, B and Smith, M L (1984) *Meta-Analysis in Social Research* Sage Publications, Beverly Hills

80   Gerbarg, Z B and Horwitz, R I (1988) 'Resolving Conflicting Clinical Trials - Guidelines for Meta-Analysis' *Journal of Clinical Epidemiology* vol 41, pp503–509

81   Needleman, N and Gatsonis, C (1990) 'Low-Level Lead Exposure and the IQ of Children, A Meta-analysis of Modern Studies' *Journal of the American Medical Association* 2 February, vol 263, no 5, pp673–78

82   Gerborg and Horwitz op cit, p676

83   Needleman and Gatsonis, op cit

84   ibid

85   Needleman and Gatsonis, op cit, p676

86   Needleman and Gatsonis, op cit, p677

87   ibid

88   ibid

89   Yule and Rutter (1985) pp241-2

90   ibid

91   Winneke, G et al (1990) 'Results from the European Multicentre Study on Lead Neurotoxicity in Children: Implications for a Risk Assessment' *Neurotoxicology and Teratology* vol 12, pp553-9

92   ibid

93   Winneke et al op cit, p556

94   Winneke et al op cit, p556-7

95   Winneke et al op cit, p557

96   World Health Organisation's International Programme on Chemical Safety (1995) *Inorganic Lead* Environmental Health Criteria 165, Geneva

97   Needleman and Gatsonis (1990) p677

98   ibid

99   ILZRO (1991) *Comments on Health and Toxicology Sections of the OECD Lead Document* Appendix I, p7

100 ibid

101 ibid

102 Ernhart, C B (1992) 'A critical review of low-level prenatal lead exposure in the human: 2. Effects on the developing child' *Reproductive Toxicology* vol 6, p22

103 When discussing the Bayle MDI scale in 1988, the MRC observed that such scales '... were not originally designed to be sensitive to small and subtle differences in performance between children [of different ages]' MRC, 1988, p19, para 5.13

104 Shaheen, S (1984) 'Neuromaturation and behavior development: the case of childhood lead poisoning' *Developmental Psychology* vol 20, pp542-50

105 ibid

106 Baghurst, P A et al (1992) 'Environmental exposure to lead and children's intelligence at the age of seven years' *New England Journal of Medicine* 29 October, vol 327, No 18, pp1279-84

107 *Australian/International Meeting on Non-Occupational Exposures to Lead: Meeting Summary* (October 1992) Melbourne, p2

108 ibid pp2-3

109 ATSDR (October 1991) *Toxicological Profile for Lead*, draft for public comment prepared by the Clement International Corporation, p49

110 Bellinger, B et al (1987) 'Longitudinal analyses of prenatal and postnatal lead exposure and early cognitive development' *New England Journal of Medicine* 23 April, vol 316, no 17, p1038

111 ATSDR (1991) op cit, p49

112 Bellinger et al (1987) p1038

113 Gradient Corporation (1990) *The relationship between blood lead and cognitive development scores: evidence from prospective epidemiological studies* 10 December, Cambridge MA, p8

114 Bellinger, D et al (1990) 'Antecedents and correlates of improved cognitive performance in children exposed in utero to low levels of lead' *Environmental Health Perspectives* vol 89, pp5–11; Bellinger, D et al (1991) 'Low-Level Lead Exposure, and Children's Cognitive Function in the Preschool Years' *Pediatrics* February, vol 87, no 2, pp219–27; cf Gradient Corp, op cit, p4

115 Bellinger, D et al (1991) 'Low-Level Lead Exposure, and Children's Cognitive Function in the Preschool Years' *Pediatrics* February, vol 87, no 2, p222, Table 2

116 ILZRO (1991) *Comments on Health and Toxicology Sections of the OECD Lead Document*, August, p6

117 Bellinger, D et al, (1992) 'Low-level lead exposure, intelligence and academic achievement: a long-term follow-up study' *Pediatrics* December, pp855–861

118 Bellinger et al op cit, p855

119 ibid

120 ibid

121 Bellinger et al op cit, pp858–9

122 Bellinger et al op cit, p859

123 Dietrich, K N et al (1990) 'Lead exposure and neurobehavioral development in late infancy' *Environmental Health Perspectives* vol 89, p14; and Gradient Corp, p5

124 Dietrich et al op cit, pp14–15

125 Dietrich et al op cit, p14

126 Dietrich et al op cit, p15

127 ibid

128 ibid

129 Dietrich et al op cit, p16

130 Dietrich et al op cit, p17

131 Dietrich, K N et al (1991) 'Lead exposure and the cognitive development of urban preschool children: the Cincinnati lead study cohort at age 4 years' *Neurotoxicology and Teratology* vol 13, no 2, pp203–11

132 Dietrich et al op cit, p203

133 Dietrich, K N et al (1992) 'Lead exposure and the central auditory processing abilities and cognitive development of urban

preschool children: the Cincinnati lead study cohort at age 5 years' *Neurotoxicology and Teratology* vol 14, no 1, pp51–56

134 Dietrich et al op cit, p51

135 Mahaffey, K R (1992) 'Exposure to Lead in Childhood' *The New England Journal of Medicine* 29 October, vol 327, no 18, pp1308–9; referring to Dietrich, K N et al (1993) 'The Developmental Consequences of Low to Moderate Prenatal and Postnatal Lead-Exposure – Intellectual Attainment in the Cincinnati Lead Study Cohort Following School Entry' *Neurotoxicology and Teratology* vol 15, no1, pp37–44

136 Mahaffey, op cit, p1309

137 Dietrich, K N et al (1993) 'The Developmental Consequences of Low to Moderate Prenatal and Postnatal Lead-Exposure – Intellectual Attainment in the Cincinnati Lead Study Cohort Following School Entry' *Neurotoxicology and Teratology* vol 15, no 1, p42

138 Ernhart, C B (1992) 'A critical review of low-level prenatal lead exposure in the human: 2. Effects on the developing child' *Reproductive Toxicology* vol 6, p28, Table 1

139 ibid

140 Gradient Corp, op cit, p6

141 Ernhart, C B (1992) 'A critical review of low-level prenatal lead exposure in the human: 2. Effects on the developing child' *Reproductive Toxicology* vol 6, p28, Table 1; Gradient Corp, op cit, p6

142 Gradient Corp, op cit, p3

143 Greene, T and Ernhart, C B (1993) 'Dentine Lead and Intelligence Prior to School Entry: A Statistical Sensitivity Analysis' *Journal of Clinical Epidemiology* vol 46, no 4, pp323–39

144 Greene, T and Ernhart, C B (1993) 'Dentine Lead and Intelligence Prior to School Entry: A Statistical Sensitivity Analysis' *Journal of Clinical Epidemiology* vol 46, no 4, pp323–39

145 Greene and Ernhart op cit, p11

146 Greene and Ernhart op cit, p14

147 ibid

148 World Health Organisation's International Programme on Chemical Safety (1995) *Inorganic Lead*, Environmental Health Criteria 165, Geneva, pp176 and 184

149 *Australian/International Meeting on Non-Occupational Exposures to Lead: Meeting Summary* (October 1992) Melbourne, Table 1, p27

150 McMichael, A J et al (1988) 'Port Pirie cohort study: environmental exposure to lead and children's abilities at the age of four years' *New England Journal of Medicine* 25 August, vol 319, no 8, p469

151 Baghurst, P A et al (1992) 'Environmental exposure to lead and

children's intelligence at the age of seven years' *New England Journal of Medicine* 29 October, vol 327, no 18, p1279

152 ibid

153 ibid

154 Wigg, N R (1988) 'Port Pirie cohort study: childhood blood lead and neuropsychological development' *Journal of Epidemiology and Community Health* vol 42, p215

155 Wigg op cit, p218

156 McMichael, A J et al (1988) 'Port Pirie cohort study: environmental exposure to lead and children's abilities at the age of four years' *New England Journal of Medicine* 25 August, vol 319, no 8, pp468–75

157 Baghurst et al (1992) p1280

158 Baghurst et al op cit, p1281

159 Baghurst et al op cit, p1282

160 Tong, S et al (1996) 'Lifetime exposure to environmental lead and children's intelligence at 11–13 years: the Port Pirie cohort study' *British Medical Journal* 22 June, vol 313, pp1569–75

161 Tong et al op cit, p1573

162 Tong et al op cit, p1574

163 ibid

164 Tong et al op cit, p1574

165 Cooney, G H et al (1989) 'Low-level exposure to lead: the Sydney lead study' *Developmental Medicine and Child Neurology* vol 31, p640

166 Gradient Corp, op cit, p7

167 Cooney, G H et al (1989) 'Low-level exposure to lead: the Sydney lead study' *Developmental Medicine and Child Neurology* vol 31, p641

168 ibid

169 ibid

170 Cooney, G H et al (1989) 'Neurobehavioural consequences of prenatal low level exposures to lead' *Neurotoxicology and Teratology* vol 11, p97

171 Cooney, G H et al (1989) 'Low-level exposure to lead: the Sydney lead study' *Developmental Medicine and Child Neurology* vol 31, pp643–4

172 cf Gradient Corp, op cit, pp7 and 11

173 Cooney, G H et al (1989) 'Neurobehavioural consequences of prenatal low level exposures to lead' *Neurotoxicology and Teratology*, p101

174 World Health Organisation's International Programme on Chemical Safety (1995) *Inorganic Lead*, Environmental Health Criteria 165, Geneva, p181, Table 21

175 ibid

176 World Health Organisation's International Programme on Chemical Safety (1995) *Inorganic Lead*, Environmental Health Criteria 165, Geneva

177 ibid p181, Table 21

178 World Health Organisation's International Programme on Chemical Safety (1995) *Inorganic Lead*, Environmental Health Criteria 165, Geneva, p224

179 Rosen, J F (1992) 'Effects of Low Levels of Lead Exposure' *Science*, 17 April, p294

180 Royal Commission on Environmental Pollution (1983) *Lead in the Environment* Ninth Report, HMSO, London, p137, para 8.19

Notes to Chapter 4

1   Royal Commission on Environmental Pollution (1983) *Lead in the Environment*, Ninth Report, HMSO, London, p136

2   Patterson, C et al (1991) 'Natural skeletal levels of lead in *homo sapiens sapiens* uncontaminated by technological lead' *Science of The Total Environment* vol 107, pp205–36

3   OECD (draft text dated 12 October 1992) *Risk Reduction Strategy Document for Lead* p95, Fig 55

4   US EPA (1991) *Users Guide for Lead: A PC Software Application of the Uptake/Biokinetic Model*, Version 0.50, January

5   Annest, J L et al (1983) 'Chronological trends in the blood lead levels between 1976 and 1980' *New England Journal of Medicine* vol 308, No 23, pp1371–7; Annest, J L et al (1983) 'Trends in the blood lead levels in the United States population: The second national health and nutrition examination survey (NHANES II) 1976–1980' in M Rutter and R Jones (eds) *Lead versus Health* John Wiley and Sons, pp33–59

6   Pirkle, J L et al, (1994) 'The decline in blood lead levels in the United States' *Journal of the American Medical Association*, 27 July vol 272 No 4, pp284–91

7   Mahaffey, K R et al (1982) 'National estimates of blood levels: United States, 1976–1980: associated with selected demographic and socioeconomic factors' *New England Journal of Medicine* vol 307, p577

8   Mahaffey et al op cit, p98

9   ibid

10 Centre for Disease Control (1978) 'Preventing lead poisoning in young children' *Journal of Pediatrics* vol 93, No 4, pp709–20

11 K R Mahaffey et al (1982) 'National estimates of blood levels: United States, 1976–1980: associated with selected demographic

and socioeconomic factors' *New England Journal of Medicine* vol 307, p576

12   Mahaffey et al op cit, p577, Table 3

13   Mahaffey et al op cit, p576

14   ATSDR (1988) *The Nature and Extent of Lead Poisoning in Children in the United States*, Department of Health and Human Service, Atlanta

15   ATSDR op cit, pV–6

16   ATSDR op cit, pI–47

17   ATSDR op cit, p4

18   ATSDR op cit, p I–48

19   *Mortality and Morbidity Weekly Report* 95–10–27, CDC, Atlanta

20   Binder, S and Falk, H (1991) *Strategic Plan for the Elimination of Childhood Lead Poisoning* CDC, Atlanta; EPA (1991) *Strategy for Reducing Lead Exposures*, February 21

21   Brody, D J et al (1994) 'Blood lead levels in the US population' *Journal of the American Medical Association*, 27 July vol 272 No 4, pp277–83; J L Pirkle et al (1994) 'The decline in blood lead levels in the United States' *Journal of the American Medical Association*, 27 July vol 272 No 4, pp284–91

22   Brody, D J et al (1994) 'Blood lead levels in the US population' *Journal of the American Medical Association*, 27 July vol 272 No 4, p277

23   Pirkle, J L et al (1994) 'The decline in blood lead levels in the United States' *Journal of the American Medical Association*, 27 July vol 272 No 4, pp284–91

24   Brody, D J et al (1994) 'Blood lead levels in the US population' *Journal of the American Medical Association*, 27 July vol 272 No 4, pp277–83

25   Norman, E H and Clayton Bordley, W (1995) 'Lead toxicity intervention in children' *Journal of the Royal Society of Medicine*, March vol 88, pp121–4

26   Norman and Clayton Bordley op cit, p122

27   Royal Commission on Environmental Pollution (1983) *Lead in the Environment*, Ninth Report, HMSO, London, para 4.34

28   Sherlock, J C and Quinn, M J (1986) 'Relationship between blood lead concentrations and dietary lead intake in infants: the Glasgow duplicate diet study 1979–1980' *Food Additives and Contaminants* vol 3, No 2, pp167–8

29   It was known that drinking water in Glasgow often contained levels of lead above 100 µg/L. Quinn, M J and Sherlock, J C (1990) Correspondence between UK 'action levels' for lead in blood and in water, *Food Additives and Contaminants* vol 7, No 3, p393

30   Sherlock, J C and Quinn, M J (1986) 'Relationship between blood lead concentrations and dietary lead intake in infants: the Glasgow duplicate diet study 1979–1980' *Food Additives and Contaminants* vol 3, No 2, p168

31   Sherlock, J C and Quinn, M J (1986) 'Relationship between blood lead concentrations and dietary lead intake in infants: the Glasgow duplicate diet study 1979–1980' *Food Additives and Contaminants* vol 3, No 2, p172 Table 2

32   Sherlock, J C and Quinn, M J (1986) 'Relationship between blood lead concentrations and dietary lead intake in infants: the Glasgow duplicate diet study 1979–1980' *Food Additives and Contaminants* vol 3, No 2, p170

33   EEC Directive 77/312/EEC *Official Journal*, L105/10–17 April 1977

34   Department of the Environment (1981) *European Community Screening Programme for Lead: United Kingdom Results for 1979–1980*, Pollution Report No 10

35   Royal Commission on Environmental Pollution (1983) *Lead in the Environment*, Ninth Report, HMSO, London, para 4.35, p48

36   Department of Health and Social Security (1980) *Lead and Health*, The Lawther Report, Appendix 2, p98

37   DHSS op cit, Table 4 5, p49

38   DHSS op cit, Figure 4 3, p49

39   DHSS op cit, Figure 4 4, p50

40   DHSS op cit, para 4 38 p51

41   ibid

42   Royal Commission on Environmental Pollution (1983) *Lead in the Environment*, Ninth Report, HMSO, London, Appendix 4, Table 1, p155

43   RCEP op cit, Chapter II

44   RCEP op cit, Para 4 40 - 4 42, esp p53

45   Zarembski, P M et al (1983) 'Lead in neonates and mothers' *Clinica Chimica Acta* vol 134, pp35–49

46   ibid

47   EEC Directive 77/312/EEC *Official Journal*, L105/10–17 April 1977

48   Department of Health and Social Security (1980) *Lead and Health*, The Lawther Report, HMSO

49   Bryce-Smith, D and Stephens, R (1981) *Lead or Health*, Conservation Society, London

50   Bryce-Smith and Stephens op cit, p12

51   Quinn, M J and Delves, H T (1985) 'UK blood lead monitoring programme: results for 1984' *Heavy Metals in the Environment* vol 1, p306

52 Quinn and Delves op cit, p307
53 Quinn, M J and Delves, H T (1989) 'The UK blood lead monitoring programme 1984–1987: results for 1986' *Human Toxicology* vol 8, pp205–20
54 Quinn, M J and Delves, H T (1985) 'UK blood lead monitoring programme: results for 1984' *Heavy Metals in the Environment* vol 1, p306
55 Quinn, M J and Delves, H T (1989) 'The UK blood lead monitoring programme 1984–1987: results for 1986' *Human Toxicology* vol 8, p206
56 ibid
57 Quinn, M J and Delves, H T (1985) 'UK blood lead monitoring programme: results for 1984' *Heavy Metals in the Environment* vol 1, p308
58 ibid
59 Quinn, M J and Delves, H T (1987) 'UK blood lead monitoring programme – preliminary results for 1986' *Heavy Metals and the Environment* vol 2, p207
60 Quinn, M J and Delves, H T (1989) 'The UK blood lead monitoring programme 1984–1987: results for 1986' *Human Toxicology* vol 8, p209
61 Davies, D J A et al (1990) 'Lead intake and blood lead in two-year-old UK urban children' *The Science of the Total Environment* vol 90, p16, Table 1
62 Davies et al op cit, p210, Table 3
63 ibid
64 Davies et al op cit, p216, Figure 5
65 Quinn, M J, Delves, H T and Davies, D J A (1989) 'UK Blood lead monitoring programme for 1987' *Heavy Metals in the Environment* vol 2 pp246–9
66 Department of the Environment *UK Blood Lead Monitoring Programme 1984–1987* HMSO
67 Department of the Environment (1991) *Digest of Environmental Protection and Water Statistics*, No 14, HMSO, p15
68 Quinn, M J and Delves, H T (1989) 'The UK blood lead monitoring programme 1984–1987: results for 1986' *Human Toxicology* vol 8, p205
69 Department of the Environment (1991) *Digest of Environmental Protection and Water Statistics*, No 14, HMSO, p13
70 K Newton, *pers. comm.*, 15 August 1994
71 Delves, H T et al (1996) 'Blood lead concentrations in United Kingdom have fallen substantially since 1984' *British Medical Journal* vol 313, 5 October, p883
72 Watt, J, Thornton, I, Delves, T and Moreton, J (1994) *Lead in Drinking Water in Blackburn*, Report of a collaborative study, December, p7

73 Delves, T, Golding, J, Smith, M and Taylor, H (1996) *Blood lead levels at two and a half years*, report to the Department of the Environment

74 See eg CDC (1992) 'Surveillance of elevated blood lead levels among adults - US, 1992' *Morbidity and Mortality Weekly Report*, 1 May vol 41, No 1

75 Russell, J and Dr A Wadge, Principle Scientific Officer of the Department of Health, *Pers. Comms*. 20 December 1993 and with T Delves on Monday 1st August 1994

76 J Russell and I House of the National Poisons Unit on Wednesday Pers comm 10 November 1993

77 I House of the National Poisons Unit *Pers. comm*. on Wednesday 10 November 1993

78 CDC (1992) 'Surveillance of elevated blood lead levels among adults – US, 1992' *Morbidity and Mortality Weekly Report*, 1 May vol 41, No17, p287

79 US EPA (1994) *Integrated Exposure Uptake Biokinetic Model For Lead in Children*, (IEUBK) Version 0.99d, February; US EPA (1994) *Guidance Manual for Integrated Exposure Uptake Biokinetic Model For Lead in Children*, February

80 Thornton, I et al (1990) 'Lead exposure in young children from dust and soil in the United Kingdom' *Environmental Health Perspectives* vol 89, pp55–60

81 EPA (1986) *Reducing Lead in Drinking Water: A Benefit Analysis*, December, pI–5 to I–7

82 ATSDR (1988) op cit, VI–35

83 EPA (1986) *Reducing Lead in Drinking Water: A Benefit Analysis*, December, pI.7

84 ATSDR (1988) *The Nature and Extent of Lead Poisoning in Children in the United States*, Department of Health and Human Service, Atlanta, pVI–38

85 Patterson, J W (1981) *Corrosion in water distribution systems*, report to the EPA, Office of Drinking Water; cf EPA (1986) *Reducing Lead in Drinking Water: A Benefit Analysis*, December, pI–6, and pII–17 ff

86 EPA op cit pII–22

87 ATSDR (1988) *The Nature and Extent of Lead Poisoning in Children in the United States*, Department of Health and Human Service, Atlanta, pVI–36

88 ATSDR op cit, pVI–37

89 ATSDR (1988) op cit, pVI–37

90 ATSDR op cit, pVI–40; referring to EPA (1986) *Reducing Lead in Drinking Water: A Benefit Analysis*, Document No EPA–230–09–86–019

91 ATSDR (1988) pVI–40

92 ibid

93 ibid

94 Jeff Cohen at EPA, *pers. comm.* 29 November 1995

95 CDC (1991) *Preventing Lead Poisoning in Young Children: A Statement by the Center for Disease Control*, October, p18

96 US Department of Housing and Urban Development (1990) *Comprehensive and Workable Plan for the Abatement of Lead-Based Paint in Privately Owned Housing – A Report to Congress*, December, pp1–1 to 1–2

97 Rabin, R (1989) 'Warnings unheeded: a history of childhood lead poisoning' *American Journal of Public Health* vol 79, No 12, pp1668–74, see esp p1670

98 US HUD op cit

99 ibid

100 ibid

101 ATSDR (1988) pVI–13

102 ibid

103 ATSDR op cit, pVI–14

104 ATSDR (1988) pVI–14; referring to Pope, A (1986) *Exposure of children to lead-based paints*, Report to EPA Strategies and Air Standards Division, EPA contract no 68–02–4329; US Bureau of the Census (1986) *American housing survey 1983*, Part B: Indicators of housing and neighborhood quality by financial characteristics, December

105 US Department of Housing and Urban Development (1990) *Comprehensive and Workable Plan for the Abatement of Lead-Based Paint in Privately Owned Housing – A Report to Congress*, 7 December

106 HUD op cit, ppxvii–xviii

107 HUD op cit, pVI–29

108 ATSDR (1988) op cit

109 ibid

110 ibid

111 Weitzman, M et al (1993) 'Lead-contaminated soil abatement and urban children's blood lead levels' *Journal of the American Medical Association*, 7 April vol 269 No 13, pp1647–54

112 US EPA (1994) *Urban Soil Abatement Demonstration Project* vol 1: Integrated Report,

113 ATSDR (1988) pVI–19

114 ATSDR op cit, pVI–24

115 EPA (1991) *National Air Quality and Emission Trends Report*, 1990 p3–29

116 EPA (1991) *National Air Quality and Emission Trends Report*, 1990, p3–30

117 EPA op cit, 3–29

118 EPA op cit, 4–19

119 EPA op cit, p4–7. NB The EPA make it clear that this does not mean that every individual who lives in a violating county is exposed to levels above the NAAQS

120 ATSDR (1988) op cit, pVI–44

121 Quinn, M J and Sherlock, J C (1990) 'Correspondence between UK "action levels" for lead in blood and in water' *Food Additives and Contaminants*, 1990 vol 7, No 3, p411

122 Burke, T (1982) *Lead in Drinking Water*, Water Research Centre, Stevenage, p7

123 Christison, R (1844) 'On the action of water upon lead' *Transactions of the Royal Society of Scotland* vol 15, pp265–76

124 Quinn, M J and Sherlock, J C (1990) 'Correspondence between UK "action levels" for lead in blood and in water' *Food Additives and Contaminants*, 1990 vol 7, No 3, p411; referring to Department of the Environment (1976) *Lead in Drinking Water: a Survey in Great Britain, 1975–76*, Pollution paper No 12, HMSO

125 Sherlock, J C and Quinn, M J (1986) 'Relationship between blood lead concentrations and dietary lead intake in infants: the Glasgow duplicate diet study 1979–1980' *Food Additives and Contaminants* vol 3, No 2, p171, Figure 1

126 Quinn, M J and Sherlock, J C (1990) 'Correspondence between UK "action levels" for lead in blood and in water' *Food Additives and Contaminants* vol 7, No 3, p410

127 Moore, M R et al (1977) 'Contribution of lead in drinking water to blood lead' *Lancet*, No 8039, 24 Sept, pp661–2

128 MAFF (1982) *Survey of Lead in Food: Second Supplementary Report*, 10th Report of the Steering Group on Food Surveillance, Food Surveillance Paper No 10, p13, para 46

129 Department of the Environment (1977) *Lead in Drinking Water. A survey in Great Britain 1975–1976: Pollution Paper No 12*, HMSO, London; cf Royal Commission on Environmental Pollution (1983) *Lead in the Environment*, Ninth Report HMSO, London, p18, para 2.13, Table 2.5

130 ibid

131 Royal Commission on Environmental Pollution, *Lead in the Environment*, Ninth Report, 1983, HMSO, London, p68

132 Burke, T (1982) *Lead in Drinking Water*, Water Research Centre, Stevenage, p11

133 ibid

134 op cit

135 Sherlock, J C et al (1984) 'Reduction in exposure to lead from drinking water and its effect on blood lead concentrations', *Human Toxicology* vol 3, pp383–92

136 Royal Commission on Environmental Pollution (1983) *Lead in the Environment*, Ninth Report, HMSO, London, para 4.38

137 Sherlock, J C et al (1984) 'Reduction in exposure to lead from drinking water and its effect on blood lead concentrations', *Human Toxicology* Table 2, p386

138 Sherlock, J C et al (1984) 'Reduction in exposure to lead from drinking water and its effect on blood lead concentrations', *Human Toxicology* Table 2, p386

139 Sherlock et al op cit, p386 and Table 3, p387

140 Sherlock, J C et al (1984) 'Reduction in exposure to lead from drinking water and its effect on blood lead concentrations', *Human Toxicology* vol 3, p389

141 Beveridge, J M and Graydon, R W 'Renfrew District – lead pipe survey' *Journal of the Royal Environmental Health Institute of Scotland* vol 2, No 11, pp1–4

142 Beveridge and Graydon op cit, p2

143 Beveridge and Graydon op cit, p3

144 Royal Commission on Environmental Pollution (1983) *Lead in the Environment*, Ninth Report, HMSO, London, p137, para 8.24

145 Quinn, M J and Sherlock, J C (1990) 'Correspondence between UK "action levels" for lead in blood and in water' *Food Additives and Contaminants* vol 7, No 3, p411

146 Quinn, M J and Sherlock, J C (1990) 'Correspondence between UK "action levels" for lead in blood and in water' *Food Additives and Contaminants* vol 7, No 3, p412

147 Dowling, S (1992) 'Statement by the Chief Environmental Health Officer for Blackburn Borough Council' made on *Checkout (Programme No 6)*, Channel 4 Television, 1 July; cf Watt, J, Thornton, I, Delves T and Moreton, J (1994) *Lead in Drinking Water in Blackburn*, Report of a collaborative study, December, p1 and Table 1 p15

148 North West Water (1992) *Water & Health*, June, p4

149 Watt, J, Thornton, I, Delves, T and Moreton, J (1994) *Lead in Drinking Water in Blackburn*, Report of a collaborative study, December, Tables 1 and 2

150 Watt et al op cit, p6

151 F White, Quality Compliance Manager, North West Water, *pers. comm.* 10 September 1996

152 White op cit

153 Sharp, C *pers. comm.* 7 August 1996

154 Watt, J, Thornton, I, Delves, T and Moreton, J (1994) *Lead in Drinking Water in Blackburn*, Report of a collaborative study, December, p1

155 Quinn, M J and Sherlock, J C (1990) 'Correspondence between UK "action levels" for lead in blood and in water' *Food Additives and Contaminants* vol 7, No 3, p408

156 Chambers, V K and Hitchmough, M D (1992) *Economics of lead pipe replacement (TMU 9030), Final Report, the Department of the Environment*, prepared by the Water Research Centre for the Department of the Environment, document No DoE 2956-/1, May

157 House of Lords Select Committee on the European Communities (1996) *Drinking Water, Session 1995–96* 4th Report, 30 January, p9

158 House of Lords Select Committee on the European Communities (1996) *Drinking Water, Session 1995–96* 4th Report, 30 January, p9

159 ibid

160 Quinn, M J and Sherlock, J C (1990) 'Correspondence between UK "action levels" for lead in blood and in water' *Food Additives and Contaminants* vol 7, No 3, p408

161 Gibson, J L (1904) 'A plea for painted railings and painted walls of rooms as the source of lead poisoning amongst Queensland children' *Australian Medical Gazette* vol 23, pp149–53 cited by Rabin, R (1989) 'Warnings unheeded: a history of childhood lead poisoning' *American Journal of Public Health* vol 79, No 12, pp1668–74

162 Horner, J M (1994) 'Lead poisoning from paint – still a potential problem' *Journal of the Royal Society of Health* vol 114, No 5, October, pp245–247

163 Housing and Urban Development (1990) *Comprehensive and Workable Plan for the Abatement of Lead-Based Paint in Privately Owned Housing – A Report to Congress*, 7 December; Housing and Urban Development, The HUD Lead-Based Paint Abatement Demonstration (FHA), August; Farfel, M R et al (1991) 'An evaluation of experimental practices for abatement of residential lead-based paint: report on a pilot project' *Environmental Research* vol 55, pp199–212; Farfel, M R et al (1994) 'The longer-term effectiveness of residential lead paint abatement' *Environmental Research* vol 66, pp217–21

164 Housing and Urban Development (1990) *Comprehensive and Workable Plan for the Abatement of Lead-Based Paint in Privately Owned Housing – A Report to Congress*, 7 December; Housing and Urban Development (1991) The HUD Lead-Based Paint Abatement Demonstration (FHA), August

165 US EPA (1991) *Strategy for Reducing Lead Exposures*, 21 February; US EPA (1992) *Summary: Lead Strategy Implementation Status*, Office of Pollution Prevention and Toxics, 17 June

166 *Lead and Health* (1980) DHSS, p34, para 69

167 O'Neill, P (1992) 'Lead paint: still a threat to children' *British Medical Journal*, 22 August vol 305, p440

168 Royal Commission on Environmental Pollution (1983) *Lead in the*

Environment, Ninth Report HMSO, London, p137, para 8.26; cf Sherlock et al (1984) 'Reduction in exposure to lead from drinking water and its effect on blood lead concentrations', *Human Toxicology* vol 3, pp387–91

169 Royal Commission on Environmental Pollution (1983) *Lead in the Environment*, Ninth Report, HMSO, London, p142

170 Department of the Environment (1983) *Lead in the Environment*, Pollution Paper No 19, p10

171 Department of Trade and Industry, *pers. comm.* 16 December 1996

172 DoE op cit, p11

173 Thornton I et al (1990) 'Lead Exposure in Young Children from Dust and Soil in the UK' *Environmental Health Perspectives* vol 89, pp55–60

174 Thornton, I et al (1990) 'Lead Exposure in Young Children from Dust and Soil in the UK' *Environmental Health Perspectives* vol 89, pp55–60

175 Thornton et al op cit, p58

176 Dr J Watt of Imperial College, *pers. comm.*, Friday 19 November, 1993 with J Russell

177 J Atherton, Department of the Environment, *pers. comm.* 14 March 1995

178 Davies, D J A et al (1990) 'Lead intake and blood lead in two-year-old UK urban children' *The Science of the Total Environment* vol 90, pp13–29, see esp p17; cf Thornton, I et al (1990) 'Lead Exposure in Young Children from Dust and Soil in the UK' *Environmental Health Perspectives* vol 89, p59

179 Thornton, I et al (1990) 'Lead Exposure in Young Children from Dust and Soil in the UK' *Environmental Health Perspectives* vol 89, p59

180 Steenhout, A (1987) 'Kinetic and epidemiologic approaches of the human organism responses to lead exposure: a solution for the Pb-blood/Pb-air, or Pb-water, Pb-dust… relationships' *Heavy Metals in the Environment*, pp280–82

181 Thornton, I et al (1990) 'Lead Exposure in Young Children from Dust and Soil in the UK' *Environmental Health Perspectives* vol 89, p59

182 Russell, J, *pers. comm.*, 16 May 1996

183 Ashley, A Environmental Health Officer, Brighton Borough Council, *pers. comm.*, 14 June 1995, with J Russell

184 O'Neill, P (1992) 'Lead paint: still a threat to children' *British Medical Journal*, 22 August vol 305, p440

185 Nedellec, V et al (1995) 'Evaluation des travaux de decontamination de 59 logements d'enfants atteints de saturnisme' *Revue D'Epidemiologie et Sante Publique* vol 43, No 5, pp485–93

186 Department of the Environment)1991) *Digest of Environmental Protection and Water Statistics* No 14; and Russell, J and Millstone, E 'The Reduction of Lead Emissions in the UK' (1995) in M Jänicke & H Weidner (eds), *Successful Environmental Policy: a critical evaluation of 24 cases*, Rainer Bohn, Berlin

187 Haigh, N (1992) *Manual of Environmental Policy: The EC and Britain*, Institute of European Environmental Policy/Longmans, Section: 6 7–7

188 ibid

189 Beasley, M (1983) *Scientific, Technical, Economic and Political Factors in the Controversy over Lead in Petrol*, Sussex University MSc Dissertation at the Science Policy Research Unit, p9

190 *The Guardian*, 29 June 1984, p8

191 See Section 8.25 in The Ninth Report of the Royal Commission on Environmental Pollution, HMSO, 1983

192 Cawse, P A et al (1987) 'Trace and major elements in the atmosphere at rural locations in Great Britain 1972–81' in P Coughtrey, M H Martin and M H Unsworth (eds), *Pollution transport and the fate of ecosystems*, Blackwell, Oxford, pp89–112

193 Department of the Environment (1991) *Digest of Environmental Protection and Water Statistics*, No 14, HMSO, p15

194 National Society for Clean Air (1988) *Unleaded Petrol – What's the Problem?*, 31 October, p1

195 ibid

196 Haigh, N op cit, Section 6 7–6

197 Russell, J and Millstone, E (1995) 'The Reduction of Lead Emissions in the UK' in M Jänicke & H Weidner (eds), *Successful Environmental Policy: a critical evaluation of 24 cases*, Rainer Bohn, Berlin, p231, Figure 3

198 Department of the Environment (1995) *Digest of Environmental Statistics* p25, Table 2.15; McDanell, R (1996) Department of the Environment, *pers. comm.*, 16 December

199 Monier-Williams, G M (1938) *Lead in Food*, Reports of Public Health and Medical Subjects No 88, Ministry of Health, HMSO

200 MAFF (1982) *Survey of Lead in Food: Second Supplementary Report*, 10th Report of the Steering Group on Food Surveillance, Food Surveillance Paper No 10, HMSO, London, p18

201 ibid

202 MAFF (1982) *Survey of Lead in Food: Second Supplementary Report*, 10th Report of the Steering Group on Food Surveillance, Food Surveillance Paper No 10, p14, para 50

203 Department of the Environment: Central Directorate on Environmental Pollution (1983) *Lead in the Environment: The Government's General Response*, Government response to the

Ninth Report of the Royal Commission on Environmental Pollution, Pollution Paper No 19, p7

204 MAFF (1982) *Survey of Lead in Food: Second Supplementary Report*, 10th Report of the Steering Group on Food Surveillance, Food Surveillance Paper No 10, p48, Table 11

205 ibid

206 ibid

207 Lampitt, L H and Rooke, H S (1933) 'Occurrence and origin of lead in canned sardines' *Analyst* vol 58, pp733–8

208 MAFF (1982) *Survey of Lead in Food: Second Supplementary Report*, 10th Report of the Steering Group on Food Surveillance, Food Surveillance Paper No 10, p17

209 MAFF (1982) *Survey of Lead in Food: Second Supplementary Report*, 10th Report of the Steering Group on Food Surveillance, Food Surveillance Paper No 10, p7, para 27

210 ibid

211 ibid

212 Meah, M N, Smart, G A, Harrison, A J and Sherlock, J C (1991) 'Lead and tin in canned foods: results of the UK survey 1983–1987' *Food Additives and Contaminants*, vol 8, No 4, p485

213 MAFF (1982) *Survey of Lead in Food: Second Supplementary Report*, 10th Report of the Steering Group on Food Surveillance, Food Surveillance Paper No 10

214 MAFF op cit, p2 para 8 and p3 para 9

215 MAFF op cit, p32

216 Department of Health and Social Security (1980) *Lead and Health*, HMSO, p10, para 10

217 MAFF (1982) *Survey of Lead in Food: Second Supplementary Report*, 10th Report of the Steering Group on Food Surveillance, Food Surveillance Paper No 10, p5, para 19

218 MAFF op cit, p6 para 20

219 MAFF (1989) *Lead in Food: Progress Report* 27th Report of the Steering Group on Food Surveillance, Third Supplementary Report on Lead, HMSO, London

220 MAFF op cit, p1

221 MAFF op cit, p7

222 MAFF, Food Surveillance Information Sheet No 34, July 1994, '1991 Total Diet Study' Table II

223 The London Food Commission (1988) *Food Adulteration and how to beat it*, Unwin Hyman, London

Notes to Chapter 5

1 US EPA (1991) *Strategy for Reducing Lead Exposures*, Washington DC, February 21

2 US EPA (1991) *Strategy for Reducing Lead Exposures*, Washington DC, February 21, p1

3 Goldman, L R and Carra, J (1994) 'Childhood lead poisoning in 1994' *Journal of the American Medical Association*, 27 July vol 272 No 4, pp315–6

4 Goldman, L R and Carra, J (1994) 'Childhood lead poisoning in 1994' *Journal of the American Medical Association*, 27 July vol 272 No 4, pp315–6

5 US EPA (1991) *Strategy for Reducing Lead Exposures*, p27; The *Reid Bill* (Senate Bill 729), would have had a very similar effect. See *Environmental Update*, ILZRO (1993) April, p1

6 *Lead Poisoning Act*, 15 December 1926, Clause I.1.c

7 *Lead Poisoning Act* op cit Clauses I.1.f and g

8 Department of Health and Social Security (1980) *Lead and Health*, HMSO, p7 para 2

9 DHSS op cit, p89, para 210

10 Rutter, M (1983) *The Relationship between Science and Policy Making: The Case of Lead*, paper to the Royal Institution, 2 March

11 Bryce-Smith, D and Stephens, R (1981) *Lead or Health*, Conservation Society Pollution Working Party, pp57–73

12 Bryce-Smith, D and Stephens, R (1981) *Lead or Health*, Conservation Society Pollution Working Party, p102

13 Department of Health and Social Security (1980) *Lead and Health*, 1980, HMSO, p88, para 205

14 Department of Health and Social Security (1980) *Lead and Health*, HMSO, p91, para 211

15 DHSS op cit, p91 para 2

16 DHSS op cit, pp91–2

17 DHSS op cit, p20 para 44

18 Department of the Environment and Welsh Office (1982) *Lead in the Environment*, Joint Circular 22/82, 7 September

19 DoE and Welsh Office op cit, p1

20 Department of the Environment and Welsh Office (1982) *Lead in the Environment*, Joint Circular 22/82, 7 September, p2

21 ibid

22 Department of the Environment (1990) *This Common Inheritance*, HMSO, p22

23 CDC (1993) 'State activities for prevention of lead poisoning among children – US, 1992', *Journal of the American Medical Association*, 7 April vol 269, No 13, p1614

24   ibid

25   CDC (1993) 'State activities for prevention of lead poisoning among children – US, 1992', *Journal of the American Medical Association*, 7 April vol 269, No 13, p1614

26   *Pers. comm.* at EPA, in Washington DC, 13 September 1996 with B Cooke and D Cantor

27   Berlin, C M et al (1995) 'Treatment Guidelines For Lead-Exposure In Children' The American Academy of Pediatrics' Committee on Drugs, *Pediatrics* vol 96, No 1, 1995, pp155–60

28   Schwartz, J (1993) 'Beyond LOEL's, p Values, and Vote Counting: Methods for Looking at the Shapes and Strengths of Association' *NeuroToxicology*, vol 14, Nos 2–3, pp237–246

29   Quinn, M J and Sherlock, J C (1990) 'Correspondence between UK 'action levels' for lead in blood and in water', *Food Additives and Contaminants*, vol 7, No 3, p389

30   Council Directive on biological screening of the population for lead, *Official Journal of the European Communities*, L105/10–17, 1977

31   ibid

32   Quinn, M J and Sherlock, J C (1990) 'Correspondence between UK "action levels" for lead in blood and in water', *Food Additives and Contaminants*, vol 7, No 3, p389

33   Quinn and Sherlock op cit, p390

34   Pocock, S J, Smith, M and Baghurst, P (1994) 'Environmental lead and children's intelligence: a systematic review of the epidemiological evidence' *British Medical Journal* vol 309, 5 November, pp1189–97

35   Department of the Environment and Welsh Office (1982) *Lead in the Environment*, Joint Circular 22/82, 7 September, p2

36   ibid

37   Medical Research Council (1984) *The Neuropsychological Effects of Lead in Children: A Review of Recent Research 1979–1983*; Medical Research Council (1988) *The Neuropsychological Effects of Lead in Children: A Review of Recent Research 1984–1988*

38   Medical Research Council (1984) *The Neuropsychological Effects of Lead in Children: A Review of Recent Research 1979–1983*, p19

39   Medical Research Council (1988) *The Neuropsychological Effects of Lead in Children: A Review of Recent Research 1984–1988*, p19, paras 63–64

40   Pocock, S J, Smith, M and Baghurst, P (1994) 'Environmental lead and children's intelligence: a systematic review of the epidemiological evidence' *British Medical Journal* vol 309, 5 November, pp1189–97

41   Pocock et al op cit, p1189

42   ibid

43 O'Neill, P (1992) 'Lead paint: still a threat to children' *British Medical Journal*, 22 August vol 305, p440

44 US EPA (1991) *Strategy for Reducing Lead Exposures*, February 21, p23

45 ibid

46 Quinn, M J and Sherlock, J C (1990) 'Correspondence between UK 'action levels' for lead in blood and in water', *Food Additives and Contaminants*, vol 7, No 3, p392

47 Quinn and Sherlock op cit, pp392–3

48 EPA (1991) *Strategy for Reducing Lead Exposures*, February 21, p23

49 EPA (1991) 'Maximum Contaminant Level Goals and National Primary Drinking Water Regulations for Lead and Copper: Final Rule' *Federal Register*, June 7, (56 FR 26460)

50 *Federal Register*, June 7 1991, 56 FR 26460

51 ibid

52 US EPA (1996) Fact *Sheet: Lead and Copper Rule Proposed Minor Revision*, EPA Office of Water document number 812-F-96–001; cf *Federal Register*, April 12

53 ibid

54 Lead Industries Association (1991) *Comments of the LIA Inc in response to the ANPR of May 1991 re Comprehensive Review of Lead in the Environment under TSCA*, 12 August, p5

55 OECD (1992), Ch 5

56 OECD (1992), Ch 5

57 *Federal Register* (1990) *Consumer Product Safety Commission*, 55 FR 22387 to 22390, June 1; cf OECD, 1992, p151

58 Quinn, M J and Sherlock, J C (1990) 'Correspondence between UK "action levels" for lead in blood and in water', *Food Additives and Contaminants*, vol 7, No 3, p391

59 ibid

60 Quinn and Sherlock op cit, p391

61 EEC Potable Water Directive, 80/778; see parameter 51 in Annex 1 Table D

62 Quinn, M J and Sherlock, J C (1990) 'Correspondence between UK "action levels" for lead in blood and in water', *Food Additives and Contaminants*, vol 7, No 3, p391

63 ibid

64 Quinn and Sherlock op cit, p392

65 Department of the Environment (1982) *Lead in the Environment*, Circular 22/82, 7 September

66 Quinn, M J and Sherlock, J C (1990) 'Correspondence between UK "action levels" for lead in blood and in water', *Food Additives and Contaminants*, vol 7, No 3, p392; The 1989 regulations also

required a water company to replace its part of a lead pipe whenever a householder removes their part of the lead piping, and requests the water company to remove its share

67 OFWAT National Customer Council (1995) *Achieving a tighter water standard for lead in drinking water in England and Wales*, Office of Water Services, June para 22

68 op cit

69 Burke, T (1982) *Lead in Drinking Water*, Water Research Centre, Stevenage, p1

70 Department of the Environment and Welsh Office (1982) *Lead in the Environment*, Joint Circular 22/82, 7 September, Annex C, p13

71 Department of the Environment and Welsh Office (1982) *Lead in the Environment*, Joint Circular 22/82, 7 September, Annex C, p13

72 Burke, T (1982) *Lead in Drinking Water*, Water Research Centre, Stevenage, p14

73 Southern Water Services (1993) *Lead plumbing: the facts*, Worthing, January

74 Quinn, M J and Sherlock, J C (1990) 'Correspondence between UK "action levels" for lead in blood and in water', *Food Additives and Contaminants*, vol 7, No 3, p392

75 Burke, T (1982) *Lead in Drinking Water*, Water Research Centre, Stevenage, 1982, p1

76 OFWAT National Customer Council (1995) *Achieving a tighter water standard for lead in drinking water in England and Wales*, Office of Water Services, June para 13

77 *Official Journal* vol 38, C131 pp5–24, 30 May 1995; Amendment to the Drinking Water Directive 80/778/EEC

78 OFWAT National Customer Council (1995) *Achieving a tighter water standard for lead in drinking water in England and Wales*, Office of Water Services, June para 17

79 According to OFWAT National Customer Council (1995) *Achieving a tighter water standard for lead in drinking water in England and Wales*, Office of Water Services, June para 19

80 OFWAT National Customer Council (1995) *Achieving a tighter water standard for lead in drinking water in England and Wales*, Office of Water Services, June para 20

81 Friends of the Earth, *Press Release: Mandatory Grants for Lead Pipe Replacement Under Threat*, 21 February 1996, p3

82 Friends of the Earth, *Press Release: Mandatory Grants for Lead Pipe Replacement Under Threat*, 21 February 1996, p3

83 I am grateful to Dr Neil Ward of the University of Surrey for drawing my attention to this point

84 Royal Commission on Environmental Pollution (1983) *Lead in the Environment*, Ninth Report, HMSO, London, p69, para 6.6

85   ibid

86   ibid

87   ibid

88   Royal Commission on Environmental Pollution (1983) *Lead in the Environment*, Ninth Report, HMSO, London, p69, para 6.8

89   Chambers, V K and Hitchmough, M D (1992) *Economics of Lead Pipe Replacement*, (TMU 9030) Final Report to the Department of the Environment, May, p43

90   Chambers and Hitchmough op cit

91   *Housing Grants, Construction and Regeneration Bill*, HL Bill 79, 51/4, as amended on Report; cf Friends of the Earth (1996) *Press Release: Mandatory Grants for Lead Pipe Replacement Under Threat*, 21 February

92   *Housing Grants, Construction and Regeneration Bill*, HL Bill 79, 51/4, as amended on Report

93   US EPA (1992) *State Efforts to Promote Lead-Acid Battery Recycling*, Office of Solid Waste and Emergency Response, EPA/530-SW–91–029, NTIS No PB92–119965, January

94   Quinn & Sherlock, p393

95   ibid; Southern Water Services (1993) *Lead plumbing: the facts*, Worthing, January

96   Brown, P (1994) *The Guardian*, 'Environmental failures condemned' 11 May, p5; Friends of the Earth (1996) *Press Release: Mandatory Grants for Lead Pipe Replacement Under Threat*, 21 February, p2;

97   Friends of the Earth (1996) *Press Release: Mandatory Grants for Lead Pipe Replacement Under Threat*, 21 February

98   Binder, S and Falk, H et al (1991) US Centre for Disease Control, *Strategic Plan for the Elimination of Childhood Lead Poisoning*, February, p9

99   Gibson, J L (1904) 'A plea for painted railings and painted walls of rooms as the source of lead poisoning amongst Queensland children' *Australian Medical Gazette*, vol 23, pp149–53; cited by Rabin, R (1989) 'Warnings unheeded: a history of childhood lead poisoning' *American Journal of Public Health*, vol 79, No 12, pp1668–74

100   Doug Brugge (1995) 'Market Share Legislation: Holding the lead pigment companies accountable for their role in lead poisoning' *New Solutions*, Winter vol 5 No 2, pp74–80

101   Brugge op cit, p75

102   Minutes of the Annual Meeting of the Lead Industries Association June 5, 1934, held at the Waldorf-Astoria Hotel, New York

103   ibid

104   Lead-Based Hazard Abatement Act, HR 2922, known as the

Cardin Bill; *Proposed Amendment to the Inland Revenue Code of 1986*, HR 3737, known as the Luken Bill

105 *pers. comm.* with Jerome Smith of the LIA, 30 November 1995

106 US EPA (1991) *Strategy for Reducing Lead Exposures*, February 21, p5

107 ibid

108 ATSDR (1988) *The Nature and Extent of Lead Poisoning in Children in the US: A Report to Congress*, July

109 US Department of Housing and Urban Development (1990) *Comprehensive and Workable Plan for the Abatement of Lead-Based Paint in Privately Owned Housing – A Report to Congress*, 7 December. It is gratifying for an Englishman to find that the British government is not the only body eccentric enough to set standards in terms of units which mix together pre- and post-Napoleonic units, namely 'feet' and 'grams'; micrograms per square meter would have been so much more elegant

110 OECD (1992) Ch 5

111 See National Council of State Legislatures (1995) *Lead Poisoning Prevention: Directory of State Contacts 1995–96*, Compiled by L Gaer & D Farquar, October

112 Alliance to End Childhood Lead Poisoning (1991) *Guide to State Lead Screening Laws*, Washington DC

113 US EPA (1991) *Summary of Public Comments: Comprehensive Review of Lead in the Environment under TSCA*, November 22, prepared by the Chemical Control Division, Office of Pesticides and Toxic Substances

114 The Lead-Based Paint Hazard Reduction and Financing Task Force (1995) *Putting The Pieces Together*, Department of Housing and Urban Development, Document HUD–1547-LBP, July

115 Lead-Based Paint Hazard Reduction and Financing Task Force op cit, p2

116 Lead-Based Paint Hazard Reduction and Financing Task Force op cit, Exhibit 3.2, p67

117 *pers. comm.* Summer 1996

118 Control of Injurious Substances Regulations, 1989

119 OECD July 1992 draft Ch 6

120 O'Neill, P (1992) 'Lead paint: still a threat to children' *British Medical Journal*, 22 August vol 305, p440

121 Lead and Health, p35, Para 73

122 Royal Commission on Environmental Pollution (1983) *Lead in the Environment*, Ninth Report, HMSO, London, p72, para 6 14

123 ibid

124 RCEP op cit, p23, para 2.28, Table 2.6

125 O'Neill, P (1992) 'Lead paint: still a threat to children' *British*

Medical Journal, 22 August vol 305, p440

126 Royal Commission on Environmental Pollution (1983) *Lead in the Environment*, Ninth Report, HMSO, London, p137, para 8.26

127 Royal Commission on Environmental Pollution (1983) *Lead in the Environment*, Ninth Report, HMSO, London, p142

128 Department of the Environment (1983) *Lead in the Environment*, Pollution Paper No 19, p10

129 DoE op cit, p11

130 Horner, J M (1994) 'Lead poisoning from paint - still a potential problem' *Journal of the Royal Society of Health* vol 114, No 5, October, pp245–7

131 Thornton, I et al (1990) 'Lead Exposure in Young Children from Dust and Soil in the United Kingdom' *Environmental Health Perspectives* vol 89, p59

132 Department of the Environment and Welsh Office (1983) *Information Note on Lead in Paintwork*, November; Department of the Environment and the Scottish Development Department, the Welsh Office and the Central Office of Information (1983) *Lead in paint can be a danger to you and your children*, November

133 Department of the Environment (1983) *Lead in paint can be a danger to you and your children*, November

134 Inskip, M J (1987) 'Lead based paint in dwellings: the potential for contamination of the home environment during renovation' *Lead in the home environment*, Thornton, I and Culbard, E (eds), Science Reviews Ltd, Northwood, Ch 7, pp71–84

135 Department of the Environment and Welsh Office (1983) *Information Note on Lead in Paintwork*, November, p4

136 ibid

137 Health and Safety Executive (1990) *Health Hazards to Painters*, Sheffield, leaflet no 5/90

138 Section 92 of the 1936 Public Health Act. That legislation has subsequently been superseded by Section 82 of the Environmental Protection Act 1990

139 Department of the Environment (1982) *Lead in the Environment*, Circular 22/82, 7 September, p2

140 Information sheet on Monitoring and Research from the Institution of Environmental Health Officers, sent to J Russell, 21 August 1992

141 Ashley, A (1995) Environmental Health Officer, Brighton Borough Council, Wednesday 14 June, *pers. comm.* with J Russell; cited in Russell, J (1996) Chapter 5, fn 98

Notes to Chapter 6

1   Goldman, L R and Carra, J (1994) 'Childhood lead poisoning in
     1994' *Journal of the American Medical Association*, 27 July
     vol 272 No 4, pp315–6

2   Norman, E W and Clayton Bordley, W (1995) 'Lead toxicity
     intervention in children' *Journal of the Royal Society of Medicine*,
     March vol 88, pp121–4

3   Goldman, L R and Carra, J (1994) op cit

4   Department of Health and Social Security (1980) *Lead and
     Health* HMSO, London, p91, para 211

Index